PLASTICS IN AGRICULTURE

PLASTICS IN AGRICULTURE

by

P. DUBOIS

Formerly Director, Centre d'Étude des Matières Plastiques (CÉMP), and President, Comité International des Plastiques en Agriculture, Paris, France

Translated and Edited by

C. A. Brighton
M.B.E., B.Sc., F.P.R.I.

Formerly Chairman of the British Agricultural and Horticultural Plastics Association, and Consultant Polymer Scientist and Technologist

APPLIED SCIENCE PUBLISHERS LTD
LONDON

APPLIED SCIENCE PUBLISHERS LTD
RIPPLE ROAD, BARKING, ESSEX, ENGLAND

British Library Cataloguing in Publication Data

Dubois, P
  Plastics in agriculture.
  1. Plastics in agriculture
  I. Title II. Brighton, C A
  631   S494.5.P5

ISBN 0-85334-776-X

WITH 21 TABLES AND 68 ILLUSTRATIONS

© APPLIED SCIENCE PUBLISHERS LTD 1978

All rights reserved. No part of this publication may be reproduced, stored in a retrieval system, or transmitted in any form or by any means, electronic, mechanical, photocopying, recording, or otherwise, without the prior written permission of the publishers, Applied Science Publishers Ltd, Ripple Road, Barking, Essex, England

Printed in Great Britain by Galliard (Printers) Ltd, Great Yarmouth

This book is dedicated to Monsieur G. Dernis, formerly President of the national and international organisations of the Comité des Plastiques en Agriculture, for his continuing efforts on behalf of plasticulture; to Monsieur F. Buclon, who is now President of the Comité National des Plastiques en Agriculture; and to all the organisers and workers in the national organisations throughout the world.

Preface

The use of plastics in agriculture and horticulture (Plasticulture) has made considerable headway during the last decade and this is evident from the number of publications which have appeared, particularly in France.

However, it would appear that there is a need for a publication which summarises the more important information and is designed for use by engineers and technicians concerned with the various aspects of the subject, and by the manufacturers of polymeric materials and the suppliers of plastics products; it is hoped that this book will fulfil this need.

In addition to the specific applications of plastics, reference is made to work on the ageing and combustion of plastics, to the mechanical properties of thermosets and thermoplastics and also to packaging. Special attention is given to the effects of orientation on the mechanical properties of films and also to the various aspects of the use of pipes. The present status of standards and specifications, particularly those for films, is examined.

The illustrations have been specially selected according to their interest and in the wish to illustrate the development of the use of plastics in agriculture and horticulture throughout the world.

The references and bibliography are wide-ranging and extensive and have made use of publications such as *Guide de l'Utilisateur des Plastiques en Agriculture*, *Plasticulture*, a quarterly publication in English and French, and reports of the International Colloquia on Plastics in Agriculture.

P. DUBOIS

Editor's Note

Professor P. Dubois was formerly Director of the Centre d'Étude des Matières Plastiques and held the chair of plastics at the Conservatoire National des Arts et Métiers, Paris, which was created in 1955. It is fitting that a work concerned with the improvements in agriculture and horticulture by the use of plastics should be written by Professor Dubois, who was concerned in the creation of the International Organisation for Plastics in Agriculture and was its first President.

Most of the early work concerned with this subject has been carried out in France and the United States and it is not surprising, therefore, that the subject matter relies heavily on publications which have appeared in these two countries. In translating this book for English readers I have retained a major part of the original text, but, since its publication in France, numerous developments have appeared; many of these have arisen in attempts to reduce the energy costs of forced cultivation following the dramatic and continuing rise in the price of fuel after the oil crisis in 1973. I have included what I consider to be the most important of these developments and at the same time the bibliography has been extended to include references of particular interest to readers of English.

The fight against rampant starvation in so many parts of the world is being carried on with unremitting vigour by many UN agencies, and in promoting new techniques in agriculture and better utilisation of the local resources there is a growing appreciation of the very important role which plastics can play. It is hoped, therefore, that this book will be a useful guide to all those who are concerned with the work on improving agricultural and horticultural methods, and particularly to those readers of English, in order to help them in improving the living standards of their fellow men.

I am most grateful to Janine Herscovici, Assistant to the Secrétaire Général of the Comité des Plastiques en Agriculture; her guidance, encouragement and help are responsible in no small way for the fact that this work now appears in English. I would like to thank my friend Henry R. Spice who read the text and made many helpful comments on the revised version.

And finally for her continued patience and fortitude in typing the script, I am most grateful to Mrs M. Moore.

Contents

Preface vii
Editor's Note ix

Chapter 1
INTRODUCTION TO PLASTICULTURE

Aims and Scope of Plastics in Agriculture and Horticulture . . . 1
Materials 1
Plant Biology 1
Climates and Microclimates 6

Chapter 2
GENERAL BACKGROUND TO PLASTICULTURE

Introduction 9
Polymers and Resins 9
Plastics Compositions 10
Energy Input 17
Other Properties of Materials Used in Plasticulture . . . 28
Comparison of Properties of Organic (Polymeric) and Mineral Glasses 32

Chapter 3
ROLES OF THE PRINCIPAL MATERIALS AND PRODUCTS

Windbreaks 33
Nets 39
Films 43
Plastics Sheets, Liners and Vessels 53

Powders and Granules 54
Tubing and Piping 54
Miscellaneous Applications 63

Chapter 4
PROPERTIES OF SEMI-FINISHED PRODUCTS IN PLASTICULTURE

Coefficient of Expansion 69
Molecular Orientation and Anisotropy 69
Properties under Tensional Stress 72
Ageing and the Quality Standard 73
Contact Embrittlement of Film 75
Cutting of Films and Sheets 76
Shaping of Sheets 76
Adhesion and Gluing 77
Welding 77
Labelling 81

Chapter 5
PLASTICULTURE IN PRACTICE

Pattern of Cultivation 82
Mulching 82
Tunnel Structures for Semi-Forcing 88
Plastics Greenhouses 92
Film Greenhouses 96
Rigid Plastics Greenhouses 105
Different Systems of Heating 110
Ventilation 113
Lining of Greenhouses 115
Techniques in the Field 116

Chapter 6
STANDARDS AND SPECIFICATIONS

General 142
Standards Tests 142
Identification of Plastics 148

Chapter 7
RESULTS ACHIEVED WITH PLASTICS

Cultivation Techniques 149
Plastics in Tropical Horticulture 151
Conservation of Produce 152
Water Management 152
Economics 152
The World-Wide Development of Plasticulture 153
Conclusions 153

Chapter 8
PROSPECTS FOR PLASTICULTURE

General Situation 155
Usage of Plastics in Various Countries 155
Extension of Present Uses 159
Usage of Plastics in New Techniques 159

References and Bibliography 162
Index 171

Chapter 1

Introduction to Plasticulture

AIMS AND SCOPE OF PLASTICS IN AGRICULTURE AND HORTICULTURE

Plasticulture is the science and the technology of the use of plastics in agriculture and horticulture. It is concerned with:
(i)   the protection of plants and crops and the improvement of cultural techniques;
(ii) packaging, preservation and storage of the produce;
(iii) the problems of local engineering, particularly in water handling;
(iv) building construction and agricultural machinery.

Plasticulture is therefore the *engineering* of the use of plastics in agriculture and horticulture.

MATERIALS

The principal materials used are:
> *films* for the creation of a microclimate and the distribution of moisture in the soil, its sterilisation and the formation of reserves of water;
> *tubes and pipes* for water supply, irrigation and drainage;
> *sheets*, both solid and foamed for insulation materials, packaging and construction.

PLANT BIOLOGY

The different phases of the development of plants are concerned with: germination, growth (assimilation by the action of chlorophyll), evapotranspiration, flowering, and fruit formation. In nearly all of these stages, plastics can bring some benefits. This is particularly so with plant containers and films for soil sterilisation, for mulching and as a covering for greenhouses and protective structures.

Assimilation by the Action of Chlorophyll
The cycle of carbon by photosynthesis proceeds by the following route:

$$\text{air} \rightarrow \text{plant} \rightarrow \text{animal} \rightarrow \text{soil} \rightarrow \text{air}.$$

Plants are energy accumulators and take in carbon dioxide during the day and then return it to the air during the night. The principal synthesis of compounds consisting of carbon, oxygen and hydrogen (glucosides), can be represented by the following equation:

$$CO_2 + H_2O + (\text{chlorophyll} + \text{photons}) \rightarrow C_n(H_2O)_m + O_2 + Q \text{ cal}$$

The reaction is activated by light energy at a wavelength of 2300 Å and this occurs in the ultra-violet part of the spectrum. However, infra-red radiation also stimulates photosynthesis by the action of eight photons per molecule of chlorophyll and it should be noted that in the forcing of crops by the use of a level of carbon dioxide of six to ten times the normal, the photosynthetic activity is doubled after several weeks.

Use of Solar Energy
   Note: Green plants convert only 1% of the solar energy received at soil level during the course of photosynthesis. Chlorophyll pigments can only endure a brightness of several thousand lux, hence the need for shading during summer when the sun gives an orthogonal illumination of about 300 000 lux.

The amount of carbon dioxide in the atmosphere remains constant at a level of 0·3 parts per thousand as a result of the maintenance of the following equilibrium

$$CaCO_3 + CO_2 + H_2O \rightleftharpoons Ca(HCO_3)_2$$

between the carbon dioxide in the air and that dissolved in the waters of the sea. The carbon dioxide is restored in the air by animal and plant respiration, by decaying plant and animal matter and by combustion, but that in rocks remains unchanged although the sea deposits calcareous sediments.

There are two forms of chlorophyll α and β. The formula is

INTRODUCTION TO PLASTICULTURE

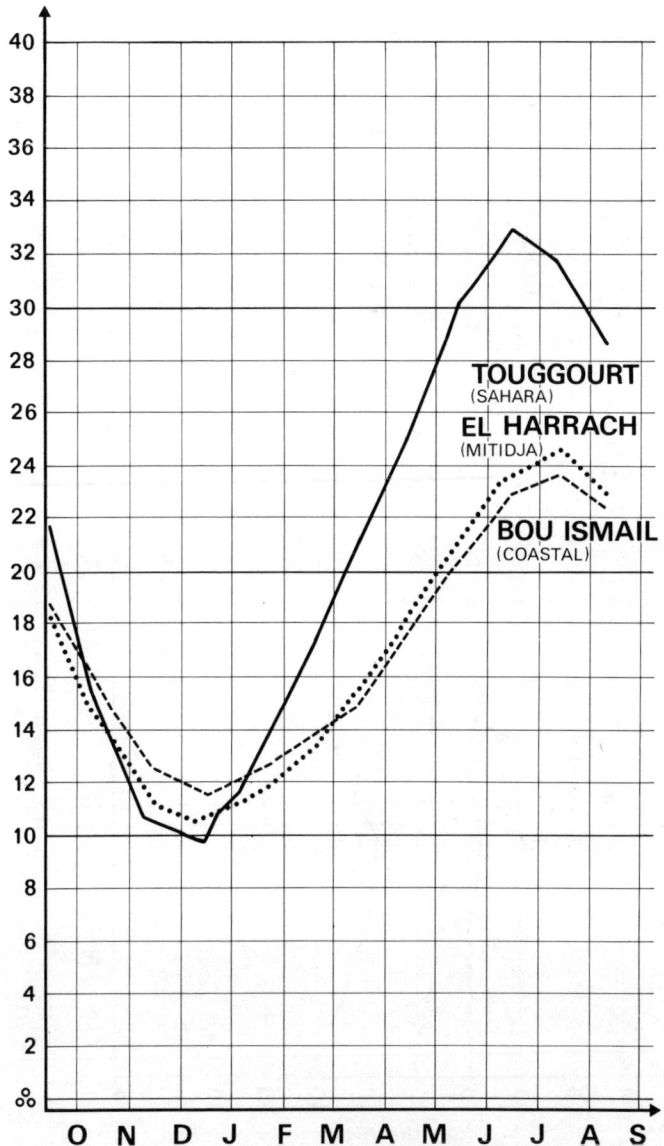

FIG. 1a. Mean temperature over a year (Algeria).[2]

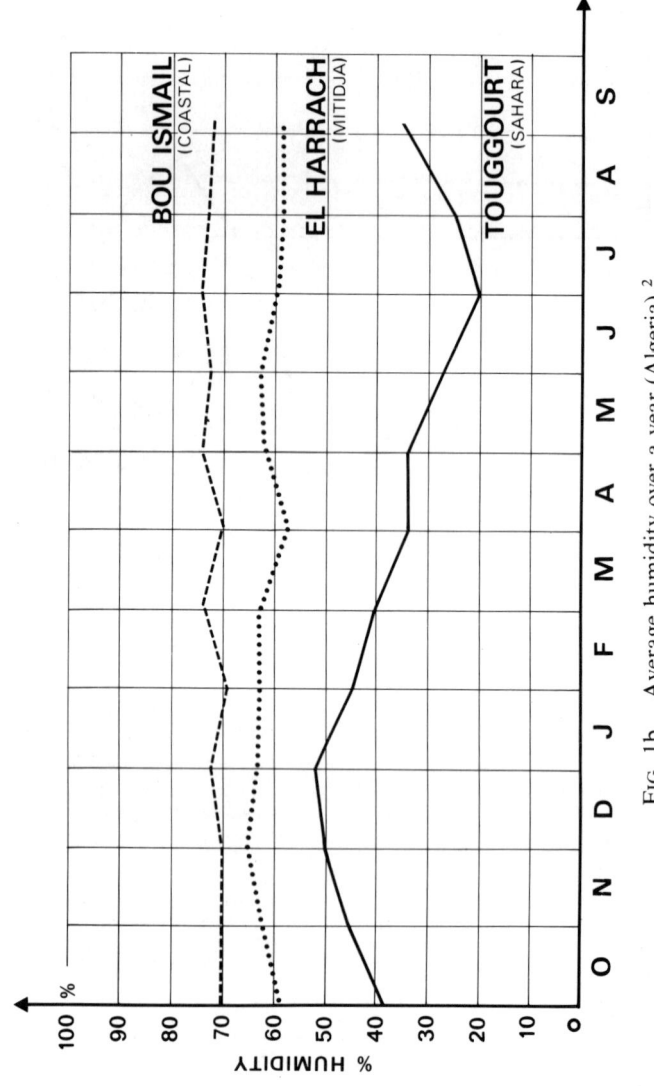

Fig. 1b. Average humidity over a year (Algeria).[2]

INTRODUCTION TO PLASTICULTURE 5

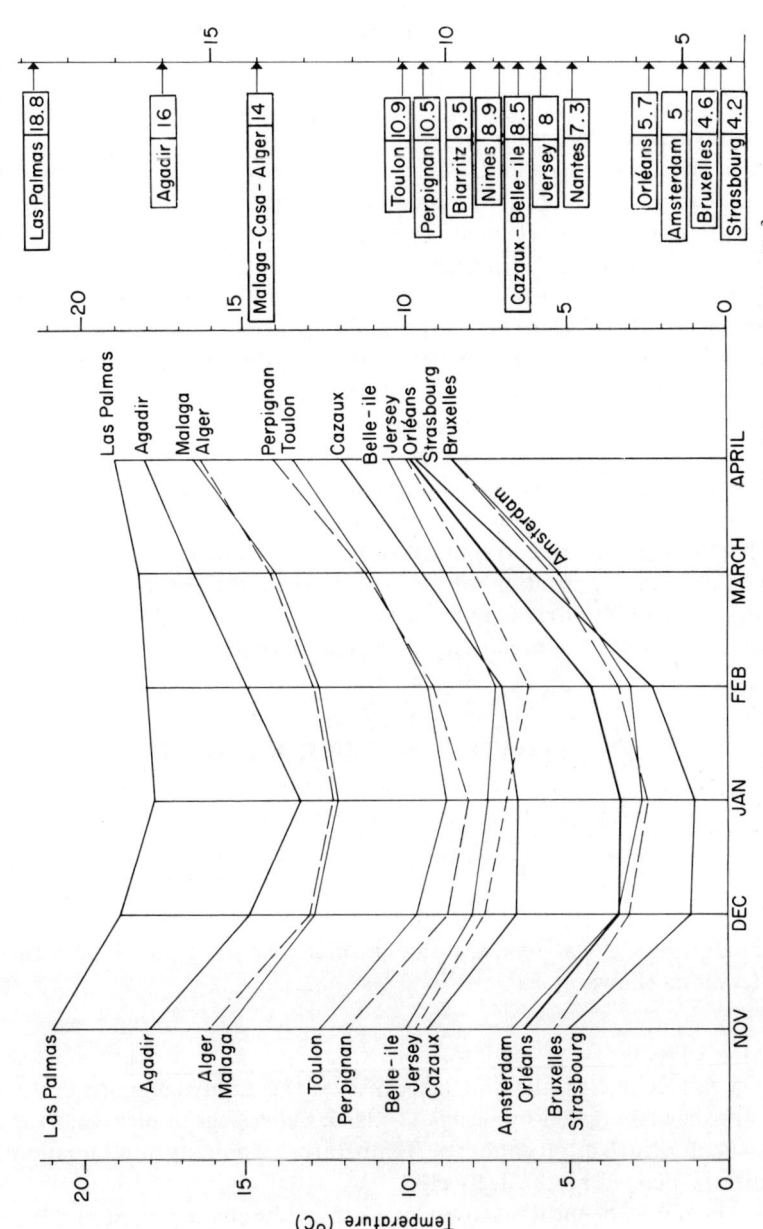

FIG. 1c. Average monthly temperatures for 6 months (November to April).[3]

$C_{55}H_{70}O_6N_4Mg$ with five heterocyclic nitrogen-containing rings, the α form containing a co-ordinated Mg atom. In the other form a methyl group,—CH_3, is replaced by an aldehyde group,—CHO.

The Nitrogen Cycle

The mineral nitrogen in the soil enters by the roots and rises by the sap to the leaves, where, under the action of sunlight, the chlorophyll promotes reaction with other elements to form compounds consisting of oxygen, nitrogen, sulphur, hydrogen and carbon (*proteins*) such as those of the plant lucerne, in which there are five sulphur bridges between the molecules. These plant proteins contain nitrogen in an organic form. They serve as foodstuffs for animals which then convert them into animal protein (meat) with accompanying losses, such as by the death of the animal or by excretion of nitrogen-containing matter. Under the action of micro-organisms, the latter is converted into ammoniacal nitrogen and then by oxidation into nitrous compounds and finally into nitric compounds. The nitrates thus complete the cycle of mineralisation. Losses occur as gas from the manure and the entrainment of the nitrates by rain-water. The gains come from the fixation of the nitrogen in the air by bacteria (e.g. *Azotobacter*, *Clostridium*) and ammonium nitrate produced by lightning during storms, manure and as nitrogenous fertilisers.

CLIMATES AND MICROCLIMATES

The course of plant biology is determined by climates and microclimates.

Types of Climate

These can be distinguished as follows: polar (Arctic and Antarctic), equatorial, tropical, temperate and insular. The classification gives further details as shown in Table 1.

Microclimates

The microclimate is the climate localised within an area of limited extent. It is artificially created by means of plastics coverings, mulch, and various structures such as greenhouses. Temperature, humidity and radiation can thus be defined more effectively.[1]

The photochemical reactivity of a site can be characterised by means of an actinometer.[2]

TABLE 1
CLIMATE CLASSIFICATION

Climate	Characteristics	Environmental factors		Others
		Temperature (°C)	Humidity (%)	
Virgin forest	Warm and humid	35–60	80–100	Moulds, termites
Swamps	Periodically dry, 1500 mm rainfall per year	12–85	45–85	Moulds, termites
Steppes	Long periods of dryness	10–45 (Daily variations can reach 23 °C)	20–70	Increased exposure to sun; dust
Deserts	Very dry, 50 mm rainfall per year	−8–50 (Daily variations can reach 50 °C)	2–30	Increased exposure to sun; sandstorms

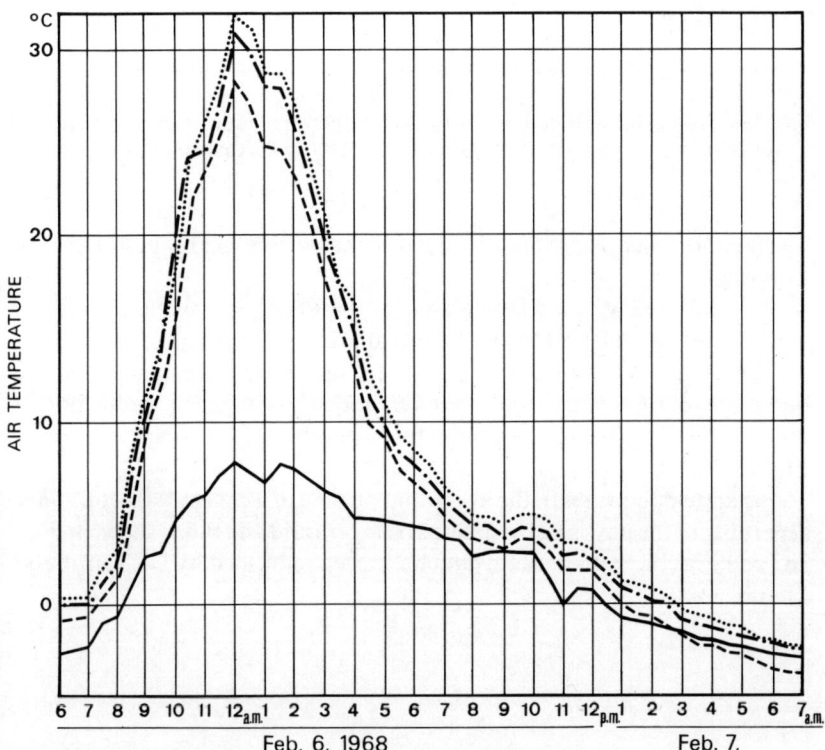

FIG. 1d. Air temperatures under PVC, EVA and polyethylene film tunnels.[4] —— outside temperature; ······ temperature under PVC film; –·–·– temperature under EVA film; – – – temperature under polyethylene film.

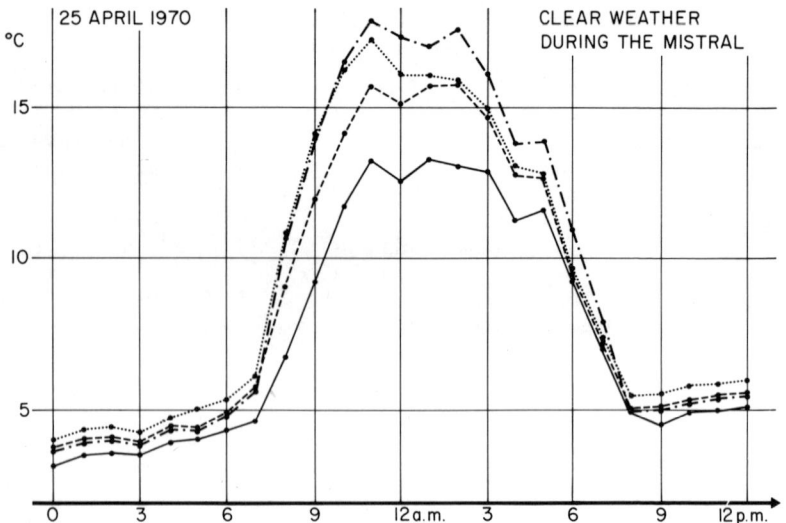

FIG. 1e. Temperature differences between the interior of tunnels and the exterior.[5] Coverings: $-\cdot-\cdot-$ anti-condensation; $---$ plain polyethylene; $-\!-\!-$ perforated polyethylene; $\cdots\cdots$ PVC.

Temperatures and Humidity of Climates and Microclimates in Relation to Covering

As shown in Fig. 1a the differences between the average outdoor temperatures at different times of the year are very pronounced, depending on the region.[2] In contrast, as Fig. 1b shows, the differences for the humidities of these same regions are very much less. A general indication of the monthly average temperatures for a number of locations is given in Fig. 1c.[3]

The air temperatures in the inside and outside of a covered structure vary according to the days and the external temperature and this is shown in Fig. 1d[4] and Fig. 1e.[5] Other useful data have been published by De Neira[6] and Manescu *et al.*[7]

Chapter 2

General Background to Plasticulture

INTRODUCTION

There is an ever increasing usage of plastics for protective structures, greenhouses, animal buildings and agricultural machinery and this is very often in association with other materials such as concrete, steel, wood and aluminium. Plastics for such applications are based on polymers or resins with the addition of a number of chemical additives such as lubricants which are the metallic salts of fatty acids (e.g. Sn or Al) and these are used at low concentrations (less than 1%) in the conversion process.

POLYMERS AND RESINS

'Pure' Polymers
In addition to the lubricant, there is normally present a low concentration of stabiliser or antioxidant (in low density polyethylene, this is less than 1% of a compound such as *p*-tertiary butyl phenol). Polymers of ethylene are based on the monomer ethylene, $CH_2 = CH_2$, which has a molecular weight M of 28; hence if the degree of polymerisation of the polyethylene is average n, i.e. the polymer is formed from n units of ethylene, then the average molecular weight M is $28n$. For a polymer which has basically a long chain structure and where the average n is 1500, the molecular weight M is 42 000.

Linear Polymers
These materials are generally made up of long chain molecules with a few short side chains; they soften under the action of heat, can be shaped under pressure and retain their form when cooled. The flexibility and solubility increase as the temperature rises. The individual molecules have a length of up to 1 μm and a width of 2–4 Å. They are coiled up and entangled into

masses which can be seen under very powerful magnification. The molecular weight and the tendency to crystallinity increase with the length of the molecular chain. These polymers are classed as thermoplastics and include such materials as polyethylene (high and low density), polyvinyl chloride (PVC), polystyrene, polypropylene, and polyamides (nylon).

Cross-linked Polymers
These are materials in which the molecular chains are connected by crosslinks, generally short, for example styrene, between the linear molecules of polyethylene glycol maleate. Such resins are capable of being shaped at low or elevated temperatures and then are cured by activators, to form a product which retains its shape on heating. These materials are classified as *thermosets* and include phenol/formaldehyde resins, urea/formaldehyde resins and polyesters.

PLASTICS COMPOSITIONS

These are based on semi-organic polymers (silicones) or organic polymers, with a number of different additives which are present to develop certain properties or to facilitate the conversion to finished products. Depending on the type of polymer, they include the following:

Reinforced Plastics
These are based either on thermosets such as phenolic, urea/formaldehyde and epoxide resins, with the incorporation of up to 30% of glass or nylon fibres and other reinforcing materials, or on *thermoplastics* such as polyethylene, polypropylene, PVC and polyamides, with similar reinforcing media.

Filled Plastics
The materials used in filled plastics can be of mineral origin, such as slate, mica or glass in powder form, or organic in nature, e.g. wood flour and cellulosic fabric, and these are incorporated in both thermosetting and thermoplastics polymers.

Plasticised Plastics
These are usually based on polyvinyl chloride which is normally rigid at ordinary temperatures and which becomes flexible with the incorporation of a plasticiser; this is normally a high boiling point organic ester such as dioctyl phthalate or tricresyl phosphate which forms a homogeneous mass with the polyvinyl chloride at elevated temperatures.

Foamed Plastics

Both thermosets and thermoplastics can be produced in a foamed form with a subsequent reduction in density. The expansion can be carried out by adding, for example, small hollow glass spheres, or a chemical agent which decomposes on heating and produces a gas (normally nitrogen) to form a porous mass, or an organic solvent such as pentane which vaporises on heating. Various techniques are used, depending on the polymer being processed, and it is possible to form foamed structures with different cell sizes, and with closed or interconnecting cells, thereby giving products with a wide range of properties. They are used for thermal insulation, packaging, buoyancy aids and light-weight structures.

Properties of Thermoplastics

The thermoplastics are the principal materials used in plasticulture, in the form of film, which is usually transparent. Their properties, particularly in relation to thermal transmission, are given by Duncan and Walker.[8] Table 2 gives a résumé of their principal properties. In the case of polyethylene, the various grades are identified by the melt flow index (MFI) which is defined by the weight in grammes of polymer extruded at 190 °C by a standard apparatus. The molten polyethylene is extruded through a die by a loaded vertical piston. The MFI varies over the range 0·2–2·0, polymers with lower values of 0·3–0·7 being required for films designed to meet the exacting agricultural standards. These particular films are characterised by their high mechanical strength and long life.

Polyvinyl chloride is the most widely used material after *polyethylene*, which dominates the agricultural and horticultural field, although *EVA*, the copolymer of ethylene and vinyl acetate, is being used in increasing quantities.

The properties of a whole range of thermoplastics in both film and rigid sheet form are given in a survey produced by R. I. Keveren of the Rubber and Plastics Research Association.[9]

Physical and Transmission Characteristics

Reference should be made to the suppliers of polymers and resins who generally have data on the mechanical properties of the various grades which they have available, but a general résumé follows. Other information has been published by Pabiot.[10]

Modulus of elasticity. This generally lies between 10 and 400 da N/mm^2 for unfilled polymers. Depending on the nature and amount of filler the modulus of a filled polymer approximates more or less to that of the filler

TABLE 2
PROPERTIES OF THERMOPLASTICS

Property	Polyamides (PA) Type 6/6 Nylon (PA 6.6)	Polyamides (PA) Type 11 Rilsan (PA 11)	Polyethylenes (PE) High density	Polyethylenes (PE) Low density	Polypropylene (PP)
Density (g/cm^3)	1·09–1·115	1·04	0·941–0·965	0·910–0·925	0·90–0·91
Light transmission	Transparent to opaque	Transparent to opaque	Translucent to opaque	Translucent to opaque	Transparent to opaque
Possibility of coloration	Unlimited	Unlimited	Unlimited	Unlimited	Unlimited
Water absorption, 24 h, thickness 3·2 mm (%)	0·4–1·5	0·5–1·0	0·01	0·015	0·01
Gas permeability of film (0·025 mm thick) at 23 °C (cm^3/m^2/24 h/1 atm)					
Oxygen	30–110	—	1600–1700	2700	1300
Nitrogen	150–390	—	440	—	—
Carbon dioxide		—	3900–10000	7700	7700
Crystallinity (%)	50	50	95	40–60	80
Elasticity modulus (kgf/mm^2)	300	150	100	10	100
Yield strength (kgf/mm^2)	4·9–7·6	4·8–66	2·5–3·9	0·7–1·4	3–4
Elongation at break (%)	90	70–300	15–100	200–575	250–700
Notched impact strength (kgf/cm/cm^2)	5·4	—	8·1–64·8	86·4	3·2–13·5
Flame propagation	Auto-extinguish	Auto-extinguish	Slow	Slow	Slow
Coefficient linear expansion (in/in/°C)	11–14·5 × 10^{-5}	11 × 10^{-5}	11–13 × 10^{-5}	16–18 × 10^{-5}	11 × 10^{-5}
Specific heat (cal/°C/g)	0·4	0·58	0·55	0·55	0·46
Deformation temperature under load (°C)	149–182	—	66–79	40–49	99–110
Resistance to continuous heat (°C)	132–159	—	121	100	135–160
Effect of sunlight	Slightly discolours	—	Carbon black essential	Surface cracks (except in brown and black)	Carbon black essential

TABLE 2—contd.

Property	Polyethylene Terephthalate (PET) (Mylar)	Rigid PVC	Flexible PVC (60% TCP)	Ethylene-vinyl acetate (EVA) 85% ethylene	Polymethyl methacrylate (PMM)	Butyl rubber
Density (g/cm^3)	1·38–1·39	1·35–1·45	1·16–1·35	0·92–0·95	1·20	1·10
Light transmission	Transparent	Transparent to opaque	Transparent to opaque	Transparent	Transparent	Opaque
Possibility of coloration	—	Unlimited	Unlimited	—	—	—
Water absorption, 24 h, thickness 3·2 mm (%)	0·5	0·007–0·4	0·15–0·75	0·05–0·13	0·35	—
Gas permeability of film (0·025 mm thick) at 23 °C (cm^3/m^2/24 h/1 atm)						
Oxygen	50	120	190–3100	7000	—	—
Nitrogen	8·4	20	53–810	—	—	—
Carbon dioxide	24	320	430–19000	2000	—	—
Crystallinity (%)	80	5–10	Amorphous	Amorphous	20	—
Elasticity modulus (kgf/mm^2)	300	300	40	57	300	0·4
Yield strength (kgf/mm^2)	11·9–18	3·5–6·3	1·05–2·45	>2	5·2	2
Elongation at break (%)	35–110	2–40	200–450	550–900	4·5	800
Notched impact strength (kgf/cm/cm^2)	—	2·2–108	Variable	No fracture	0·4	Elastomer
Flame propagation	Slow to auto-extinction	Auto-extinction	Slow to auto-extinction	—	Good	—
Coefficient linear expansion (in/in/°C)	27 × 10^{-4}	5·18–5 × 10^{-5}	7–25 × 10^{-5}	16–20 × 10^{-5}	9 × 10^{-5}	—
Specific heat (cal/°C/g)	—	0·2–0·28	0·3–0·5	0·55	0·35	—
Deformation temperature under load (°C)	—	54–74	—	35	85	—
Resistance to continuous heat (°C)	9	44–71	66–79	70	90	—
Effect of sunlight	Slightly affected	Browning	Variable according to plasticiser used	Slight	Slight	—

(1500 for wood and 7000 for glass). The opposite is the case with the addition of plasticiser, when the modulus eventually approaches that of rubber (about 0·3 da N/mm^2).

Rupture strength and elongation at break. It is generally advised that the values before and after exposure to sunlight be determined so as to follow

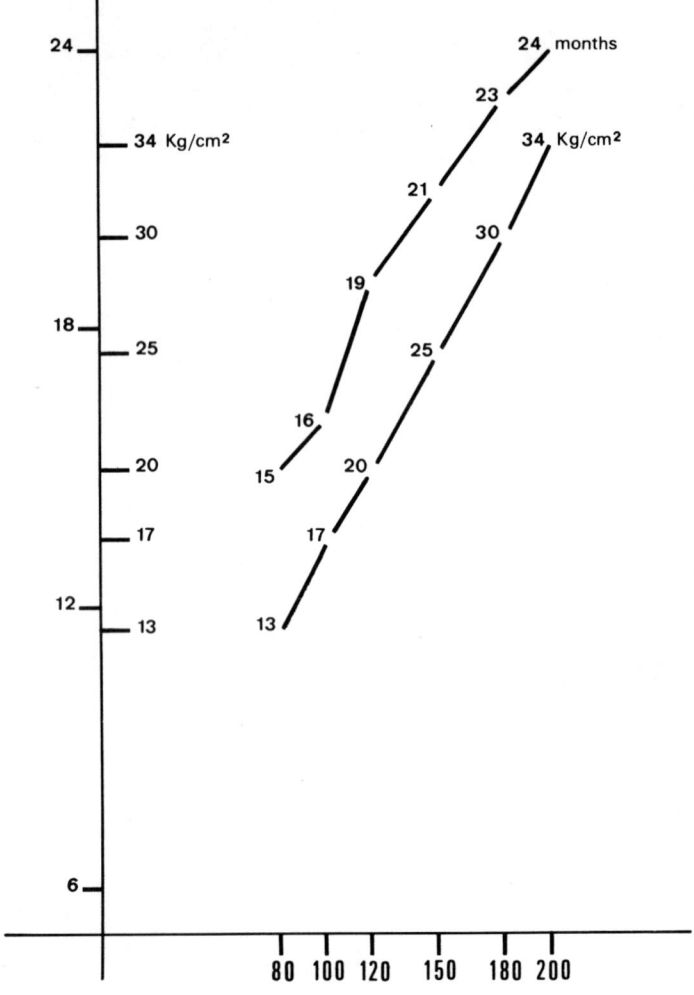

FIG. 2a. Durability of polyethylene film in relation to thickness. Ordinate: duration (months). Abscissa: thickness (μm).

the extent of ageing. The material is considered to be satisfactory until such time as the elongation falls to 50% of its original value.

(*Note:* The tensile strength of thermoplastics increases considerably with the molecular weight.)

Impact resistance. This can be determined for films by measurement of the burst strength under pressure or by the energy to failure using a falling hemispherical striker attached to a shaft onto which varying weights can be clamped (the Dart test).

Test methods. The various methods used in the physical testing of polymer films are outlined by Reid.[11]

Ageing

Ageing has received considerable attention in view of the increasing use of plastics, not only in agriculture and horticulture but also in other outside applications. The correlation of laboratory methods with ageing under natural conditions has been studied by several workers.[12-14] The factors affecting the weatherability are outlined in a number of articles[8,15-18] and can be summarised as follows:

the thickness (Fig. 2a);
the type and grade of the polymer (there are considerable differences between various grades of polyethylene as shown in Fig. 2b);

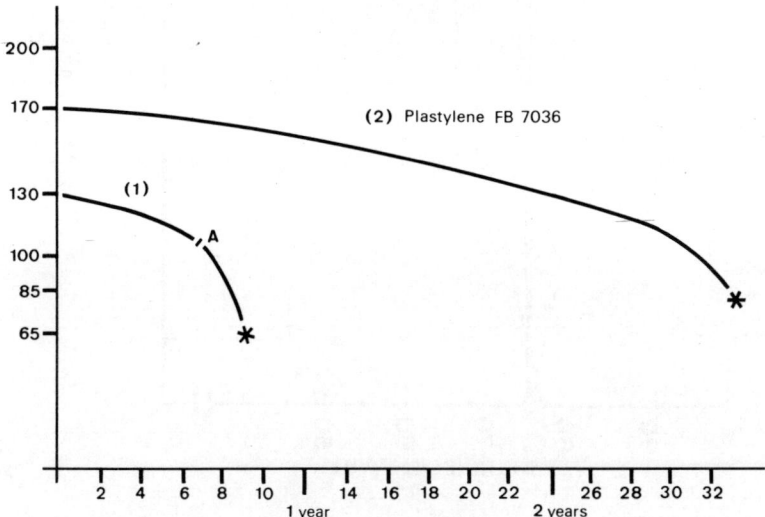

FIG. 2b. Durability in relation to nature and grade of film. Ordinate: rupture stress (kg/cm^2). Abscissa: duration (months).

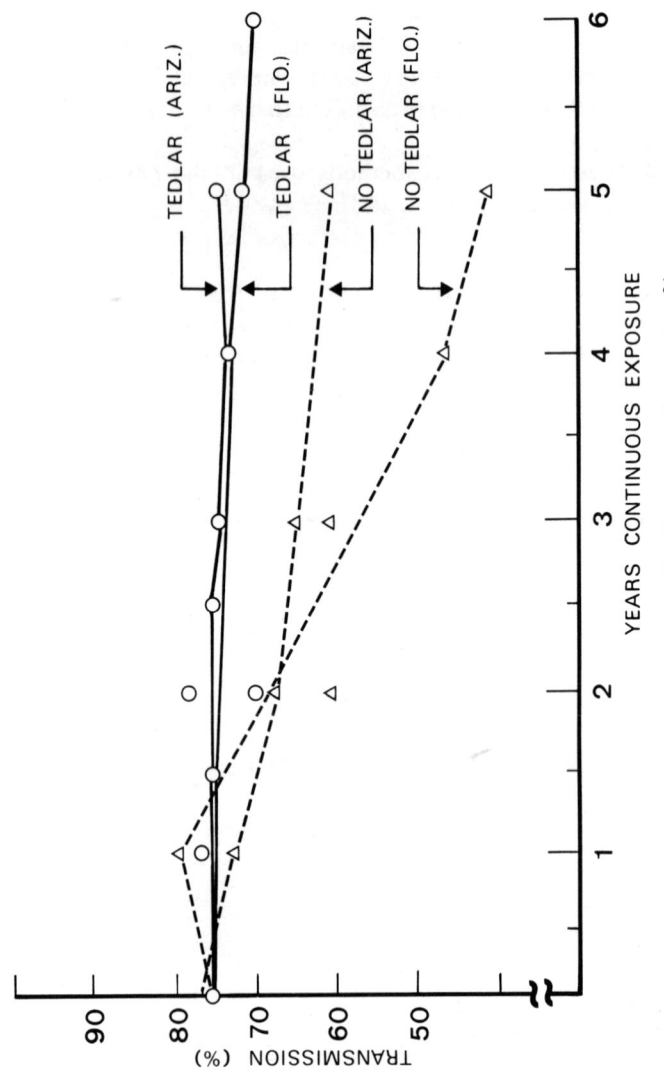

FIG. 2c. Protection of polyesters against ageing.[21]

the time of exposure: for ordinary polyethylene (curve A of Fig. 2b) the curve is sharply convex;
the ultra-violet radiation between 0·3 and 0·35 μm (although this represents only about 5% of normal sunlight);
oxygen, which participates in *photo-oxidation* thereby causing colour changes and loss of strength by embrittlement;
the temperature, which influences the degradation reactions according to the Arrhenius Law: a temperature increase of 10 °C causes a doubling or tripling of the reaction rate;
the humidity, which affects particularly fibre reinforced polyesters by surface erosion thereby exposing the fibres; this can be renewed by cleaning followed by a resin coating,[19] or can be prevented by incorporation of 15% acrylic resin in the base resin.[20] The surfacing of reinforced polyester sheets with a film of *polyvinyl fluoride* provides outstanding improvements in weathering properties[21] (Fig. 2c);
structural and minor faults (cracks, inclusion of foreign matter).

Note: Thickness plays an important role and an increase improves the durability of UV-stabilised polyethylene; for covering greenhouses and protective structures, it is recommended to use films of 180–200 μm thickness. The gain in durability is more than 20% as compared with an ordinary film for which there is no advantage in using a thicker film.

ENERGY INPUT

Solar Energy

The solar energy is the amount of energy radiated by the sun which reaches the Earth and falls perpendicularly on 1 cm² during 1 min, for the average distance of the sun from the Earth; this is 1·99 cal/min ± 0·02 = 0·13 W/cm². The intensity and the spectral distribution is affected by daily and seasonal fluctuations and by geographical and climatic variations.[22] Verdu[23] gives a distribution curve for the solar spectrum on the Earth's surface and also the average annual total energy in France. Table 3 shows the distribution observed at the Tokyorita Kogyo station in Japan.

Sunshine Hours

Naturally, the energy arriving at the soil surface depends on variations of rainfall and the geographical conditions. The map in Fig. 3[24] illustrates this for Spain.

TABLE 3
DISTRIBUTION OF DIFFERENT PORTIONS OF THE SPECTRUM DURING A FOUR-MONTH PERIOD IN 1959 AT TOKYORITA KOGYO, JAPAN[22]

Date	Average of 4 or 6 days ($mW/day/cm^2$)		
	Ultra-violet	Visible	Infra-red
August			
1–5	658	4 484	5 650
6–10	582	4 129	6 435
11–15	642	3 711	4 488
16–20	556	4 117	6 248
21–25	426	3 785	4 173
26–30	735	5 485	6 702
September			
31–4	853	5 974	7 687
5–9	548	3 964	4 566
10–14	444	3 305	3 824
15–19	662	5 117	6 006
20–24	472	3 342	4 517
25–29	480	3 094	5 049
October			
30–4	497	3 048	3 585
5–9	548	3 621	4 476
10–14	580	3 750	4 425
15–19	504	3 015	3 383
20–24	153	3 330	4 340
25–29	375	2 223	2 458
November			
30–3	379	2 034	2 390
4–9	441	2 179	2 726
9–13	244	1 223	1 353
14–18	295	2 227	3 459
19–23	328	1 848	2 750
24–28	272	1 659	2 154

Electrostatic Charges

Plastics are good electrical insulants and are widely used for this purpose. They also accumulate electric charges, which are developed by movement in contact with other materials, and these can reach such levels as to produce sparks which can lead to fires. For this reason, pipe systems through which liquids flow—particularly inflammable materials such as petrol—must be

GENERAL BACKGROUND TO PLASTICULTURE 19

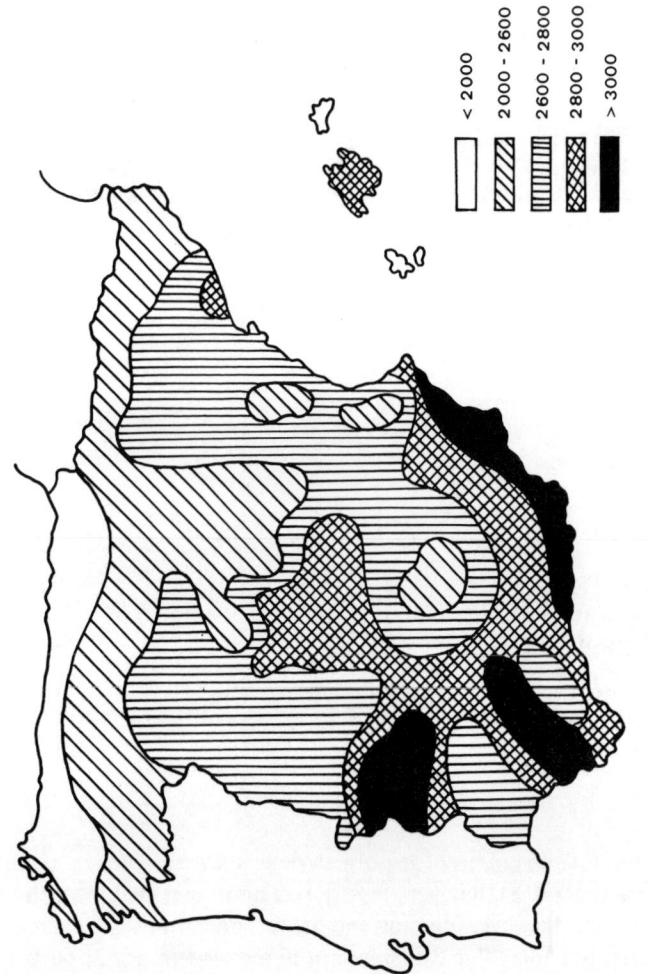

FIG. 3. Annual hours of sunshine in Spain.[24]

earthed; this is also necessary for moulding equipment. These electrostatic charges are responsible for the adhesion of dust particles, particularly from air currents in contact with plastics films, and this explains the dirt build-up on the outer surface of greenhouses and tunnels which is discussed later.

Note: Degree days. During a period of exposure over a number of days, the degree days are the sum of the average temperatures recorded in degrees centigrade.

Film Transmittance

Table 4[8] gives the relative values for single- and double-walled film structures and greenhouses; the values in parentheses are for total light transmission while the other values are for the direct transmission. The value

TABLE 4
RADIATION TRANSMISSION

Material	Solar transmission (%) Single-layer covering	Solar transmission (%) Double-layer covering	Thermal transmission (%) Single-layer covering
Polyethylene (clear)	93 (—)	88	—
Polyethylene ('Commercial clear')	76 (89)	(81)	70·8
Polyethylene (UV stabilised)	74 (88)	—	—
Glass	86 (90)	75	4·4
Transparent PVC	86 (91)	(84)	12·0
Polyvinyl 'Haze'	— (89)	(82)	—
Mylar (Polyester)	86 (90)	80	16·2
Rigid fibreglass	68 (78)	(64)	1·0

for the thermal transmission for polyethylene shown in the last column is very much higher (70·8) than for glass (4·4). This is responsible for the rapid cooling, during the evening and night, of greenhouses covered with polyethylene, but the solar transmission is reduced to about 50% of the quoted value because of the condensation of water on the internal surface.

Figure 4 shows the transmission properties before use of PVC, EVA and low density polyethylene. After two months' natural exposure transmission is reduced from 80 to 50%.[24] Table 5 summarises the reduction of physical properties after use.

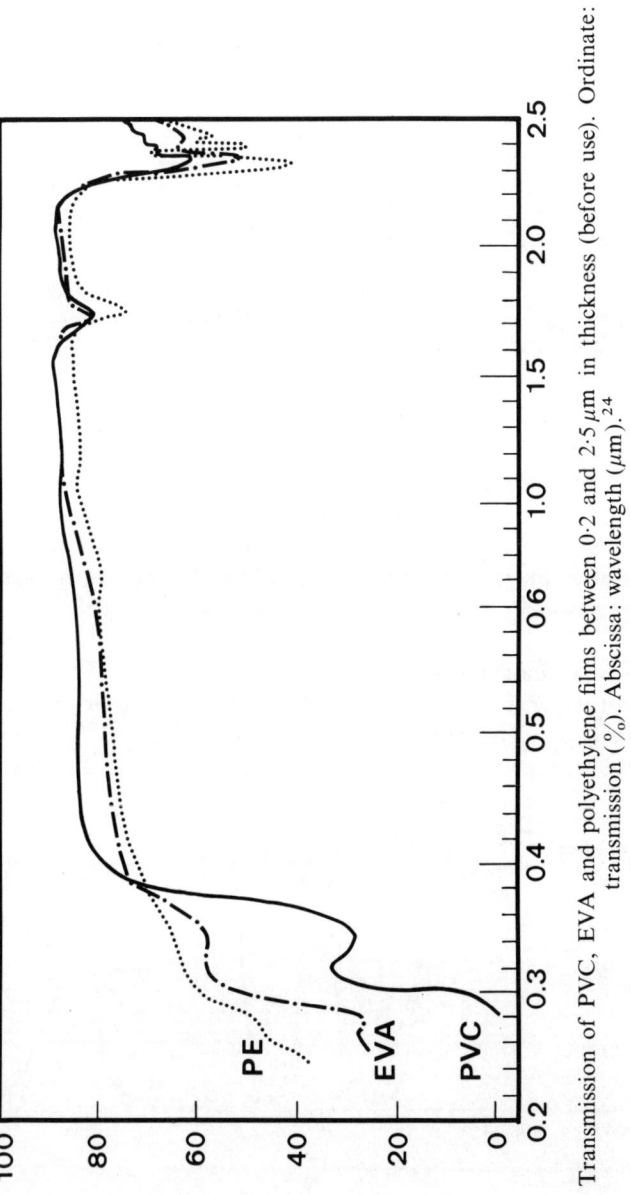

FIG. 4. Transmission of PVC, EVA and polyethylene films between 0·2 and 2·5 μm in thickness (before use). Ordinate: transmission (%). Abscissa: wavelength (μm).[24]

TABLE 5
PHYSICAL PROPERTIES OF FILMS BEFORE AND AFTER TWO MONTHS' USE

Physical property at 25°C	Before use			After use		
	PVC	EVA	PE	PVC	EVA	PE
50% Modulus (kg/cm^2)	62–73	47–59	86–90	64–75	50–64	91–95
100% Modulus (kg/cm^2)	97–116	52–71	89–96	93–119	55–78	92–101
Tensile strength (kg/cm^2)	195–234	180–196	155–164	186–232	177–198	138–158
Elongation at break (%)	250–290	517–673	493–550	238–290	460–650	446–518
Tear strength (g)	810–877	301–432	312–615	717–883	295–430	275–554
Impact load (kg/cm)	14·5	10·5	7·0	14·4	8·9	6·6

Greenhouse Effect

This arises when the solar radiation which passes through the film covering is trapped and the latter is impermeable to the infra-red radiation which is emitted in the reverse direction from the interior.

It will be seen from Fig. 5 that this effect is at a maximum when the material loses all transparency to radiation of wavelength greater than 3 μm; the emission is practically nil in the case of glass.[25]

Changes in Radiation Transmission

A drop of water on the surface of the greenhouse covering cuts out all radiation of wavelength greater than 11 μm, and about 75% between 7 and

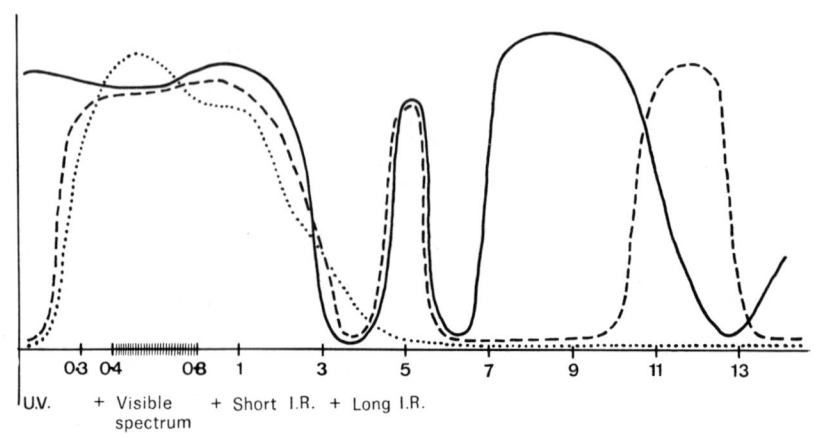

FIG. 5. Transmission spectra of polyethylene, PVC and glass. Ordinate: transparency to various wavelengths (%). Abscissa wavelength (μm): —— low density polyethylene; – – – PVC; ······ glass.

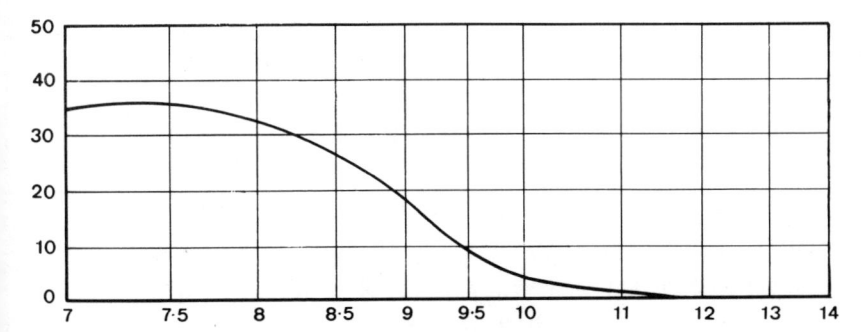

FIG. 6a. Effect of water film on spectral transmission. Ordinate: Transmission (%). Abscissa: wavelength (μm).

10 μm (Fig. 6a). It is therefore apparent that condensed water of a thickness of 0·01 mm reinforces the greenhouse effect, particularly in the case of polyethylene.

Note 1. There is, however, a small 'window' between 4 and 6 μm through which water allows the emitted radiation to pass (Fig. 5).

Note 2. The role played by water explains the interest in the spraying of a wetting agent such as 'Sun Clear' which ensures that the condensation is in

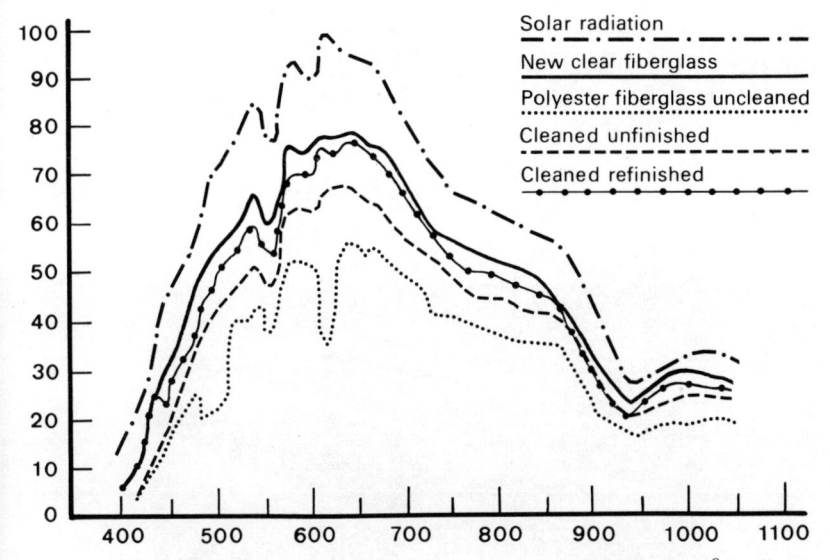

FIG. 6b. Spectral distribution and intensity of weathered GRP sheet.[8] Ordinate: spectral intensity (microwatts/cm^2). Abscissa: wavelength (nm).

the form of a continuous film instead of droplets, thereby giving better light transmission and reduced transmittance of infra-red radiation.[26]

Dirt from the atmosphere which is deposited on the outer surface of the greenhouse also reduces the light transmittance as shown in Fig. 6b, so that it becomes necessary to replace the film or to wash the covering material.[8]

Special Films

In the case of low density polyethylene, these are produced by the inclusion of different adjuvants to give the special properties.

Ultra-violet Stabilised or Long-life

This film cuts out the solar radiation of wavelengths of less than $0.3\,\mu m$ which are the most destructive since the energy decreases with the longer wavelength.[27]

Anti-mist

The film remains clear, in spite of condensation, by means of the action of a wetting agent which causes the water droplets to coalesce and run down the walls instead of dripping directly onto the plants which would encourage disease. The film allows the tips of asparagus shoots to be seen without difficulty (Fig. 7).[28]

Thermal

This type of film is modified in such a way that the radiation between the wavelengths $6-13\,\mu m$ is suppressed so that thermal polyethylene becomes comparable to PVC (Fig. 8).[25]

FIG. 7. Mulching of asparagus in Provence. (By courtesy of F. Buclon.)

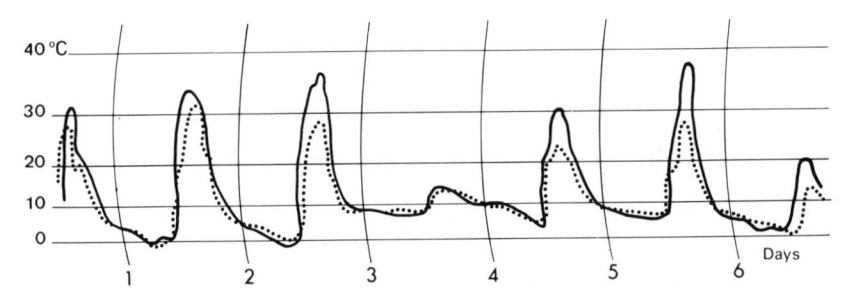

FIG. 8. Comparison of temperatures for infra-red polyethylene (——) and PVC (·····).[25]

Photo-degradable
The development of plastics, particularly polystyrene and polyethylene, with controlled degradation characteristics, was initiated in order to ensure that discarded plastics packages degrade and quickly disintegrate. Film with varying degradation rates (generally between 30 and 160 days) is available for mulching so that it breaks down after the required time and disperses into the soil. This subject was discussed at the 5th International Colloquium[29,30] and also by other authors.[31,32] The polyethylene films contain a ferric ion complex which accelerates the rate of embrittlement.[33,34] Degradation can also be promoted by incorporation of calcium carbonate into a compound based on medium or low density polyethylene. When exposed to sunlight, the material is said to disintegrate in 1–3 months and the residual inorganic materials readily disperse in the soil.[35]

Photoselectivity
Photoselectivity has been studied by several authors including Professor R. Favilli, Director of the Institute for Agronomy at Pisa,[36–38] and Glatti[39] using coloured PVC films. The effect of the inhibition|by|intense green light on the growth of weeds has been studied by Aimi.[40]

Radiation Emission by the Soil
The soil emits radiation of wavelength λt after having absorbed solar radiation of wavelength λs, according to Wien's law.

$$\lambda s \times Ts = \lambda t \times Tt = \text{a constant}$$

where T is the absolute temperature of the emitting body ($T = t\,°C + 273$). The radiation of wavelength $0.32\,\mu m$ which is given out by the sun at $6000\,°K$ is then emitted by the soil and plants at a wavelength of $7\,\mu m$ in the infra-red, assuming that their temperature is 15 °C. In a similar way, the extreme end of the solar spectrum of wavelength 2·5 μm will be absorbed and re-emitted by the plants and the soil at 54·5 μm. Consequently, in order to ensure that the solar radiation which is received in a plastics greenhouse is conserved, the film must be impermeable to the radiation from the soil between wavelengths of 7–54·5 μm in the infra-red. Glass, PVC and thermally opaque polyethylene meet these requirements.

Soil Temperature under Film

In general, under transparent film, the temperature of the soil rises by several degrees during the day; this can vary between 2–10 °C according to the season and type of soil and also according to the level of sunshine and the water content. At night, the difference in temperature between covered and uncovered soil is less (2–4 °C).

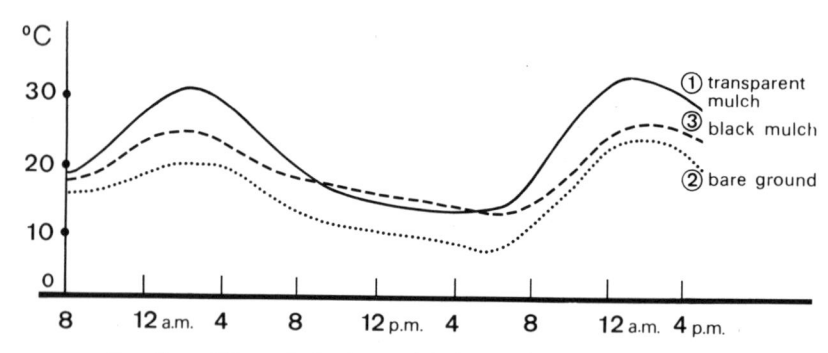

FIG. 9a. Effect of plastics mulch film on soil temperature.[41]

Under black film the soil temperature is very little higher than the control, and in fact in some instances it can even be slightly lower. The curves in Fig. 9a show the increase in temperature under mulch film[41] and in Fig. 9b the differences between various films.[42]

Under white film, the temperature is always lower than for uncovered soil. This type is used either in regions with a high level of sunshine, where it is required to reduce the transmitted radiation and soil temperature, or in regions of low luminosity, where there is a need to increase the amount of reflected light on the lower and middle leaves.

Thermal Insulation

This is characterised by the specific heat (s) and the thermal conductivity (λ)

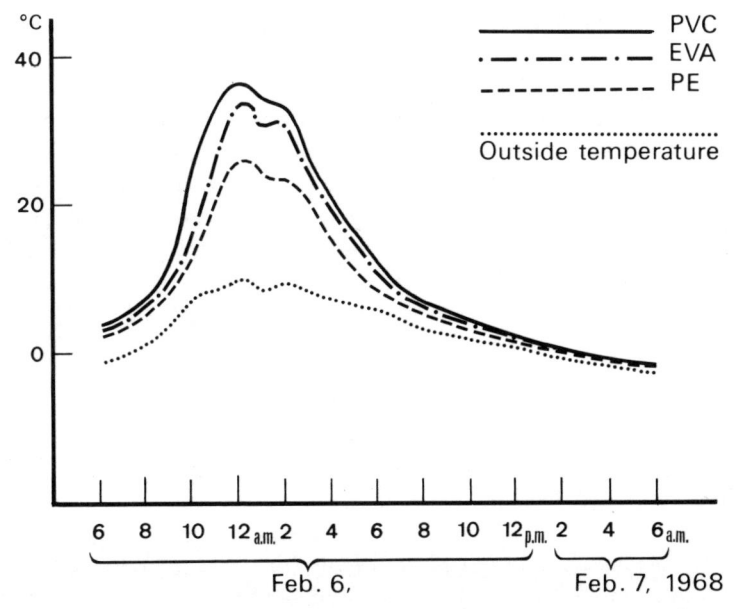

FIG. 9b. Temperature in tunnels covered with plastics films.[42]

in relation to the specific gravity (m) and the thermal diffusivity (D). The latter indicates the rate at which the temperature approaches ambient.

$$D = \frac{\lambda}{s \times m} 10^{-3} \, cm^2/sec$$

The values for certain materials are given in Table 6.

TABLE 6
THERMAL INSULATION OF CELLULAR MATERIALS

Material	Specific gravity, $m_i(g/cm^3)$	$\lambda(\times 10^{-4} \, cal/cm/sec/°C)$
Rubber	0·11	0·9
Cement	0·3	1·5
Ebonite	0·06	0·67
Phenol/formaldehyde	0·06	0·90
PVC	0·5	0·45
Polystyrene	0·08	1·0
Polyurethane	0·1	0·80
Cellular glass	0·17	1·7
Ordinary glass	2·6	20·0

Note 1: For air $\lambda = 0.58$ cal/cm/sec/°C. The value of D for ordinary glass is 37×10^{-3} cm²/sec, and this compares with about 1100 for steel which has a value of D of 950.

Note 2: Because of their good thermal insulation properties, expanded polystyrene, PVC and polyurethane are used extensively in the form of sheets and slabs. In order to prevent frost damage inside greenhouses, an insulating mat is used as a cover. For high temperature insulation, i.e. in excess of 750°C, it is preferable to use a thermorigid material such as phenol/formaldehyde foam instead of a thermoplastics type such as polystyrene.

OTHER PROPERTIES OF MATERIALS USED IN PLASTICULTURE

Combustibility and Toxicity

Because of their composition, oxyhydrocarbons (carbohydrates) produce carbon dioxide, carbon monoxide and water when burnt and the cellulosics and wood give off a certain amount of acrolein (CH$_2$=CH—CHO) which, in high concentration, causes oedema of the lungs. With the oxyazohydrocarbons (ONHC), small quantities of hydrogen cyanide are produced but the amount is much less than the quantity of carbon monoxide which causes asphyxia. Combustion of chlorine-containing plastics such as PVC gives rise to hydrogen chloride which is very corrosive to all metal objects; it attacks the respiratory system and causes concern to firemen.

Fortunately, fire hazards are very low for greenhouses and even less for unheated protective structures. The greatest risk exists in greenhouses clad with glass-reinforced polyester sheet or where wooden frames are used in agricultural buildings.

This whole subject has been examined in depth by Verdu.[43]

Note: The toxicity of vinyl chloride. Vinyl chloride is the monomer used for the manufacture of PVC. It has a boiling point of about -14°C and is therefore a gas at normal temperatures. It was thought for a long time that although the gas has anaesthetic properties, there were no long-term harmful effects. It was established in 1974 that workers engaged in the manufacture of PVC, and therefore subject to low concentrations of the gas, were liable to develop certain cancers. This has led to much more stringent precautions being taken and the amount of unchanged vinyl chloride in PVC has been drastically reduced. This is essential because of the wide-scale use of PVC for packaging foodstuffs, since many countries insist that there shall be practically no vinyl chloride migrating from the pack into the food. The polymers of PVC present no hazard when used as films and in agricultural and horticultural applications.

Permeability to Liquids and Gases

This is very low and, for water, practically zero, the absorption being of the order of 0.1 mg/m^2 after 7 days for a film of polyethylene.

As regards gases, Table 7 gives details of the permeability of a variety of materials at 25 °C for a pressure difference between the two surfaces of the film of 1 atm. Temperature has a considerable influence on the permeability of plastics films.

TABLE 7
GAS AND WATER PERMEABILITY OF FILMS AT 25 °C

Material	Water ($g/100 \text{ in}^2/0.001 \text{ in}/24 h$)	Gas ($cc/100 \text{ in}^2/0.001 \text{ in}/24 h/1 \text{ atm}$)	
		CO_2	O_2
Polyethylene			
Low density	1.0–1.5	2 700	500
High density	0.3	580	185
EVA	14	6 000	840
Polypropylene			
Cast	0.7	800	240
Oriented	0.35–0.45	370	120
Polyester	1.8	15–25	6–8
PVC			
Flexible	6–10	10–3 000	30–2 000
Rigid	4	20–50	5–20
Polystyrene	7–10	900	350
Nylon 6	—	10–12	2.6

Note 1: Alcohol vapour and perfumes readily pass through films and sheets of low density polyethylene but high density polyethylene is much less permeable because of its higher crystallinity.

Note 2: The difference in permeability to air and carbon dioxide emitted by fruit during storage is used for preservation employing a low density polyethylene pack with or without a silicone window according to the technique developed by Marcellin.[67]

TABLE 8
CHEMICAL RESISTANCE OF THERMOPLASTICS

	\multicolumn{7}{c	}{Thermoplastics}
Compound	\multicolumn{7}{c	}{Temperature (°C) below which resistance is satisfactory}
	Polymethyl-methacrylate	Polyamide Nylon 6	Polyamide Rilsan 11	Low density polyethylene	Polystyrene	Rigid PVC	Polyethylene terephthalate
A. Inorganic compounds							
(a) Elements							
Oxygen	20	20	20	20	20	60	20
Ozone	20	—	—	20	20	20	—
(b) Acids							
Hydrochloric	30% 60	0[a]	RL[b]	30% 60	30% 20	30% 60	30% 20
Nitric	10% 20	0	50% RL	10% 20	65% 20	60% R[c]	10% 20
Sulphuric	40% 20	0	40% RL	98% 20	80% 20	100% R, 98% 60	80% 0
(c) Bases							
10% Potash	60	70	R	60	20	40	0
40% Caustic soda	60	70	R	60	20	40	0
(d) Various inorganics							
Chlorine water	—	—	—	—	—	RL	—
Salt water, 30 g NaCl/litre	—	—	50	R	—	R	—
Disinfectant (based on KCl)	—	—	—	—	—	R	R
Hydrogen peroxide	—	—	—	R	—	30% R	—
B. Organic compounds							
(a) Miscellaneous							
Acetone	0	50	20	0	0	0	20
Benzene	0	20	20 Absorption	20	0	0	20
Carbon disulphide	—	—	—	0	—	—	—
Carbon tetrachloride	0	20	20 Absorption	0	0	RL	20

GENERAL BACKGROUND TO PLASTICULTURE

100% Ethyl alcohol	0	20	—	—	20	40	20
Ethyl ether	0	20	RL	—	0	0	20
Formaldehyde	20	0	—	20 Diffuses	20	40	20
Glycerol	20	90	—	60	50	60	20
Glycol	20	90	—	60	50	60	20
Methyl alcohol	0	20	20 Absorption	20 Vapour diffuses	20	40	—
Methyl ethyl ketone	0	—	—	—	0	0	—
Mineral oils	20	20	—	30	0	60	20
Petrol	—	R	R	RL	—	R	—
Phenol	—	0	0	R	—	—	—
Toluene	0	—	20 Absorption	Swells	0	0	20
Trichlorethylene	0	20	RL	—	—	—	—
White spirit	—	R	—	—	—	—	—
Xylene	0	RL	20	20	0	0	20
(b) Esters in general							
Vegetable oils	60	90	R	20	20	60	20
(c) Organic acids							
Acetic acid	60% R	10% 90	10% 90 / 50% 20 / 20	100% 20	60% 20	60% 60 / 80% R / 50	100% 20
Citric acid	—	—	—	—	—	—	—
Fatty acids	—	—	50% R	R	—	—	—
Lactic acid	20	—	100% R	—	—	—	—
Oleic acid	—	20	50% R	—	—	—	—
(d) Foodstuffs							
Alcoholic drinks	—	20	20	R	—	—	—
Butter	—	20	20	20	—	—	—
Animal fats	—	90	90	RL	—	—	—
Vegetable oils	—	90	90	RL	—	—	—
Fruit juices	—	20	20	60	—	—	—
Milk	—	20	20	60	—	—	—
Wine	—	50	50	50	—	—	—

[a] 0: no resistance; [b] RL: limited resistance; [c] R: resistance satisfactory.

Chemical Resistance of Thermoplastics

The chemical resistance of the principal materials in use is summarised in Table 8; other information is available in a publication by Verdu.[43]

TABLE 9
ENERGY REQUIRED FOR FRACTURE

Material	Energy required ($kg\,cm/cm^2$)
Window glass	0·3
Phenolic resin	
filled with wood flour	1·3
fabric-filled	4·2
Polyester fibres	80
Cast iron	2·7
Steel	200

TABLE 10
GENERAL PROPERTIES

Property	Organic glasses (Polymeric)	Mineral glasses
Density	1·2–1·5	2·3–4
Young's modulus (kgf/mm^2)	300–1 000	4 000–9 000
Rupture strength in tension (kgf/mm^2)	4–6	60–120
Rupture strength in compression (kgf/mm^2)	10–20	200
Hardness (relative values)	30–80	200
Softening temperature (°C)	50–120	450
Coefficient of expansion	$10-100 \times 10^{-6}$	$30-90 \times 10^{-7}$
Refractive index	1·45–1·80	1·49–1·75
Impact strength ($kg\,cm/cm^2$)	1–80	0·3

COMPARISON OF PROPERTIES OF ORGANIC (POLYMERIC) AND MINERAL GLASSES

Data on the energy required for fracture per unit volume (in $kg\,cm/cm^2$) of a variety of materials are listed in Table 9 and the general properties of organic and mineral glasses are summarised in Table 10.

Chapter 3

Roles of the Principal Materials and Products

WINDBREAKS

General
A windbreak is used for lowering the windspeed which, if reduced by one half, reduces the mechanical effects to about one quarter; it is thus effective as a modifier of the micro-climate, and has beneficial effects on the growth of plants.

Windbreaks have been used for a long time in certain districts (such as the valleys of the Rhone and Po), by the sea, and over wide expanses in the form of hedges, lines of trees and bamboos (Fig. 10a), and, more recently, plastics (polyethylene) netting and mesh, mounted vertically and fixed firmly between supports (Fig. 10b).[44-46] The use of other forms of plastics for this application has been reported (see, for example, Fig. 10c).[47]

Mechanism of the Functioning of Windbreaks
Although the way in which windbreaks operate is not completely understood, and studies are still being carried out, the various factors which influence the efficacy of windbreaks have been the subject of work carried out in France by Guyot.[48-50]

Figure 11 represents the effects on a flow of air when it comes up against an impermeable windbreak (A) and a permeable windbreak (B). When air meets a solid barrier it is directed upwards, and the width of the layer, as represented by the heights of the windbreak, is reduced. There is therefore an increase in air speed and this creates a reduced pressure (Bernoulli's law). The result of this is that air is drawn into the stream from downwind of the windbreak so that the air stream quickly regains its original dimensions and the static pressure increases. The air which is drawn into the stream thereby creates a *turbulent zone* immediately behind the windbreak. The

Fig. 10a. Use of bamboo canes as a windbreak.

Fig. 10b. Plastics net windbreak.

FIG. 10c. Windbreak constructed from strips of polyethylene film. (By courtesy of H. R. Spice.)

flow of the air returns to ground level fairly quickly, so the area of the so-called *protected zone* is relatively small.

With a permeable windbreak the volume of air which is deflected over the top is less; consequently the increase in speed is less and the pressure effects which lead to the formation of a turbulent zone are reduced. The deflected zone returns more slowly to its original course and hence the protected zone is longer. It has a depth of about 20 times the height of the windbreak (which is usually about 2 m).

The reduction of the windspeed gets less as the *porosity* of the windbreak increases, and it has been established that the optimum value is about 50 % porosity, although different meshes with the same porosity can have different effects on the wind. The average reduction of the windspeed is just over 50 % for an impermeable windbreak and about 40 % for the permeable type.

Effects of Height and Porosity of Windbreak
Trials have been carried out to study the behaviour of windbreaks having a porosity which is not uniform over its whole height. With a continuous plastics film (width 0·5 m) at the bottom and two different meshes above (to a total height of 2 m) the overall effect is very similar to that of an impermeable windbreak. With the reversal of the materials, the most

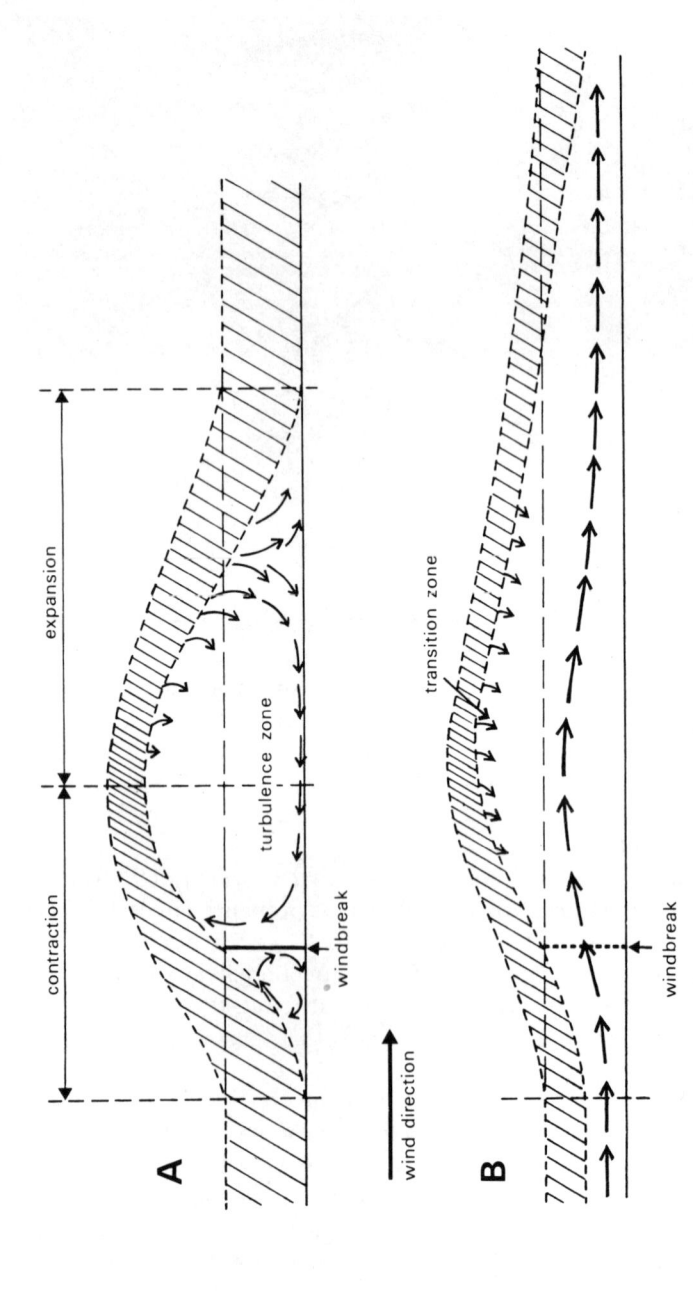

FIG. 11. Schematic representation of wind flow for solid (A) and permeable (B) windbreaks. (By courtesy of G. Guyot.)

permeable band being at the bottom, the overall effect is very similar to that of a permeable windbreak with the same porosity over its whole area.

Roughness of Ground Area
It has been established that the roughness of the ground in front of the windbreak has an effect on its efficacy; the smoother the ground, the more efficient is its action. This is also the case with the roughness of the edges of the mesh used for the windbreak.

Successive Windbreaks
A windbreak located behind another is effective over a shorter distance than the one situated upwind.

Local Climate
The protection provided by a windbreak depends in a very large measure on the thermal stability of the atmosphere. The windbreak creates an upward movement and, if the air is unstable—for example, during the day the light warm air is found near the soil surface—a large part of the incident air flow skirts the windbreak. The consequence of this is that windbreaks will have different aerodynamic effects, according to the climate of the region in which they are used. The extension of the protected zone by a permeable windbreak, which can be 20 times its height in northern climates, will be only between 12 and 15 times in Mediterranean and tropical climates.

Wind Speed
It does not appear that the protection afforded by the windbreak varies to any great extent with the wind speed. When the wind speed increases, the thermal gradient gets less and it is difficult to separate the effects due to the diminution of the thermal gradient and those due to the wind increase.

Reduction of Heat Transfer in Vertical and Horizontal Directions
Windbreaks reduce the vertical and horizontal transfer of heat. During the day, in the absence of horizontal movement, a windbreak will create a raising of the temperature of the air and of the soil. When a horizontal movement of warm or cold air is superimposed on this flow then it can result either in a lowering or in a raising of the air temperature. For example, during dry conditions, a windbreak used with an irrigated crop will result in a lowering of the temperature.

Reduction of Mass Transfer
A reduction in the transfer of water from the plant to the air produces a reduction in the potential evapotranspiration (ETP). The level of carbon dioxide in the air will also be modified.

Biological Effects of Windbreaks
Reduction of Mechanical Damage
The reduction in the speed of the wind by windbreaks is accompanied by a reduction in the amount of mechanical damage such as fall of fruit and tearing of the leaves. Spectacular results have been obtained for fruit and vegetable crops. Increases in the *area of the foliage* (an increase of 20–30% has been reported for a pear orchard) lead to better photosynthesis.[50]

Changes in Day-time Temperature
In spring the rise in temperature of the air and soil brought about by the use of windbreaks can be 1–2 °C and this encourages growth and earlier crop production. In a dry climate the temperature behind a windbreak can rise by 5 °C and this can give rise to burning and shrivelling in the absence of irrigation.

The air temperature can also fall, particularly in a warm dry climate, when the crops are irrigated; this effect of the windbreak can be useful as it creates better conditions for growth.

Lowering of Night-time Temperature
The lowering of the temperature during the night can be of the order of 1–2 °C in the presence of a windbreak and in spring this can be serious where there is a risk of frost, particularly in certain parts of France.

Modification in the Transfer of Water and CO_2
When windbreaks are used there is a reduction in the transfer of water vapour and carbon dioxide between the vegetation and the atmosphere. The reduction in the exchange of carbon dioxide is not a favourable factor but the reduction in the exchange of water vapour (due to the lowering of the ETP) is a factor which is particularly helpful to the growth of plants. This effect on the transpiration of plants explains in most cases the increased yields which have been recorded when windbreaks are used.

Considerable research has been conducted in Scotland and increases in strawberry and raspberry crop yields of 40% have been obtained. In Australia strawberry yields increased by more than 100%.[45]

Influence of Limiting Factors

The high increases reported for cucumbers and melons were for those crops where the level of production was rather low.

With other improvements which might be adopted, the increases in yield are less and it would be reasonable not to expect more than a 20% increase in those areas where winds do not take on extreme violence or where the production is already efficiently carried out.

Protection of Air-Inflated Greenhouses

The development and commercial manufacture of air-inflated greenhouses have led to a study of their resistance to wind and also to the protection of these greenhouses when they are subjected to gales. Acceptable deformations have been obtained up to a wind speed of 75 m.p.h. (120 km/h) in trials on a scale model in a wind tunnel.

Protection by means of a windbreak, reducing the wind speed to about 40 to 60% of its original value, appears necessary, the zone of protection having a height of 1–1.3 times the height of the windbreak and a length of 15 to 20 times the height.

This method can be used if necessary to protect projects and buildings in the course of construction.

Conclusions

If a primary system of protection is employed, using lines of trees spaced at 15 to 20 times their height in the direction of the prevailing wind and from 30 to 40 times in the orthogonal direction, a secondary system providing shelter for delicate subjects can be made up at stated intervals with windbreaks in plastics which can replace those constructed of the traditional canes.

The inconveniences of the traditional windbreak are not inconsiderable; there is the cost of manufacture, maintenance, and the area lost for cultivation and, in addition, it provides a habitat for certain pests such as rodents, birds and insects.

If the physical effects of windbreaks are sufficiently widely recognised, then it will be necessary to undertake a long and difficult experimental study to establish the physiological effects on different plants.

NETS

General

These are nets made from filaments of various polymers—high density polyethylene, polypropylene and polyamides—which are stretched to give the desired values of the strength and elongation at break, properties which vary according to the required usage.

Functions

Nets have essentially two roles: in the protection of plants and in crop gathering.

Protection of Plants

Physical. Protection is provided principally against bad weather, torrential rain and hail, of varying intensity according to the district, and also against birds (Fig. 12). They are used also as a means of support, for

FIG. 12. Nets for protection of lettuce against birds.

example in the growing of carnations. Sometimes the net is replaced by a continuous film.

Radiative. Nets in dark colours, most often black, are used for shading not only to provide protection against scorching but also to control photosynthesis as discussed in the opening chapter. The thermal amplitudes are reduced by 2–4 °C.

Anti-frost. Windbreaks and nets play a very important role in this aspect.

FIG. 13. Frost protection barriers.

The use of films (as shown in Fig. 13), set around the plants so as to create a channel for the flow of cold air towards the lower levels, is even more effective.

Since the air is warmer in valleys than on high ground, plants in the valleys are always more forward because of the increase in the degree-days (sum of the average temperatures for each day for the appropriate period). The opposite applies to high ground. If cold air flows down into the valley, it causes frosting of the vegetation but this does not occur on the heights where the plant growth is not so far advanced. This explains the interest in hedges and windbreaks on slopes where they increase the impedance of the descent of the cold air.

FIG. 14. Nets for collecting olives.

Crop-gathering
Nets can be used for the rapid gathering of crops such as olives and nuts, allowing the trees to be shaken without any great damage (Fig. 14).

Dimensions of Anti-Hail Nets
Anti-hail nets (Fig. 15) are normally of any required length and 4–6 m wide with a mesh size usually 4 mm × 8 mm. The net is used to form a double-slope roof supported on posts. The durability appears to be about 10 years.[51,52]

FIG. 15. Nets for protection against hail.

Net Supports for Fungicide or Insecticide
This type of net forms a protective trellis[53] which prevents mildew from the new growth stage of vines right up to leaf-fall.

Optimum Properties of Nets

Strength
This is dependent on the type of the plastics material, the stretching of the filaments during manufacture, and the resistance to ageing.

Ageing
It is clearly variable according to the duration and the season of the year, the radiative energy and the protection against ultra-violet radiation incorporated in the filaments. The least unfavourable conditions are, in principle, those in which nets are used for harvesting, which is carried out in a short time.

FILMS

General

Films play a major role in plasticulture in the creation of a micro-climate which is beneficial to growth, by virtue of their use in mulching, low tunnels, various shelters and greenhouses.

They have brought into cultivation innovations and results previously considered unattainable and have also led to an almost complete acceptance of techniques such as the mulching of melons and strawberries and of methods for the planting of new vineyards and the cultivation of maize.

There is also their use in soil sterilisation by fumigation, and also in the handling of fertilizers and their even distribution in the soil in association with water.

Functions

Mulching

Objective. This is a protective covering on the soil around plants with the aim of helping growth and, consequently, crop earliness, productivity and partial protection of the produce by suppression of weeds; protection is also provided from frost and from the action of torrential rain. Furthermore, there is a saving in water for irrigation, as well as its retention, and a saving in labour.

Mulch films and their role. Straw which has been used for mulching in the past is now often replaced by plastics film; this is generally low density polyethylene, although polyvinyl chloride (PVC) or the copolymer of ethylene and vinyl acetate (EVA) can be used. These act as follows:

on *radiation*, by using, if need be, the colour (coloured films) or the opacity obtained by metallisation (using aluminium or the oxides of titanium, aluminium, etc.);

on *soil temperature and humidity* below the film;

on *plants*, by suppression of weeds if black film is used: weed growth is normal under transparent film unless herbicides are used;

on *pests* such as gastropods and certain micro-organisms;

on *growth* in ensuring that the carbon dioxide produced by fermentation in the soil is brought into contact with the stomata of low-growing plants.

Equipment and techniques: Mulching. See Chapter 5.

General scope of usage. Mulching has been extended considerably in many countries where horticulture is highly developed. The general situation is reviewed at frequent intervals and many publications have appeared.[32,54–59]

Special types of mulch films. In addition to the black and transparent polyethylene films which are extensively used, many other types have been employed; these include black paper coated with clear polyethylene,[60,61] aluminised polyethylene, and white reflective film.[62]

The use of slit films, normally of transparent polyethylene, has been developed in the USA by Smith.[63–65] The material is cut before laying with transverse slit dimensions of 5 in wide and 1 in apart to achieve 1 in ribbons. These dimensions were chosen as the optimum after observing many combinations. These have the advantage that seedlings have no difficulty in finding their way through it and rain can also reach the plants.

Films for Packaging

Produce packing. Extensive use is now made of films for the packaging of vegetables and fruit for direct sale to the consumer. Low density polyethylene in the form of bags is very widely used: they can be purchased in relatively small quantities and in a vast range of sizes, at moderate cost. In the main they are filled and closed by hand but automatic equipment for filling and sealing is available.

The most common form of produce wrapping is that of stretch wrapping, using a highly plasticised PVC. The film is applied directly to some of the more robust items such as cabbages, or over trays, formed from expanded polystyrene sheet, containing loose produce, and as the name implies it is stretched tightly over the pack and then sealed on the underside.

The films normally used for shrink wrapping are plasticised PVC, polyethylene and polypropylene with the first-named dominating the field. The technique requires that the film is wrapped around the produce contained in a tray, and then passed through a heated tunnel when the film shrinks and holds the produce firmly in position. Oriented polyethylene is used, the degree of shrinkage depending on the amount of orientation introduced into the film at the manufacturing stage.

Produce conservation. Storage of fruit such as apples and pears may be greatly improved by using wrappings and sacks with diffusion windows. These special wrappings make use of the selective gas permeability of polyethylene films and special silicone elastomer membranes so that the fruit is kept in a controlled atmosphere, with optimum concentrations of oxygen and carbon dioxide.[66] The technique has now been developed for the bulk storage of fruit.[67]

The storage of bananas is improved by packing the fruit in polyethylene bags with an ethylene absorbent (potassium permanganate) at a temperature of about 20 °C.[68]

Small-diameter layflat polyethylene tubing (76 mm) is used to provide a compact unit for packaging of cut flower-buds. The unit can be held in cold storage for at least 8 weeks in the case of carnations. A possible 50–60% saving in transportation space is possible in comparing the number of stems per carton using tubes, with the standard method of packing mature long-stemmed flowers.[69]

Silage under Plastics Film

General. The application of plastics film to silage is of interest in a number of agricultural productions, notably corn, maize and beetroot, but especially for the conservation of green fodder crops (e.g. lucerne) which give the required increase in milk and veal production.

Principle. It is necessary to prevent as far as possible the action of the air and its humidity on the crop. Thus, only lactic acid fermentation takes place and this produces an appetising feed which is wholesome and fresh and rich in vitamins and carotene, while, in the case of acetic and butyric acid fermentations, the taste and odour are bad. The results obtained are of the following composition:

$$\text{pH} = 4\cdot 2$$
$$\text{Lactic acid} = 1\cdot 5\text{–}2\cdot 5\%$$
$$\text{Acetic acid} = 0\cdot 5\%$$
$$\text{Butyric acid} = 0\cdot 1\%$$

$$\frac{\text{Ammoniacal nitrogen}}{\text{Total nitrogen}} = 10\%$$

Note: In fact, hydrochloric, orthophosphoric and formic acids are also used in order to obtain a pH of about 4. The first two have the disadvantage of decalcifying the bones of animals but this does not happen with formic acid. On the other hand, orthophosphoric acid, like ferrous sulphate which is sometimes added, has the advantage of killing the bacteria in silage without affecting plants or animals.

Prevention of Access of Air and Water

Desiccation. This can be carried out in either of two ways:

(i) drying in the sun—this can be done either by hand or by machine;
(ii) artificial drying—this requires the expenditure of a considerable amount of energy (600 cal. per gramme of water).

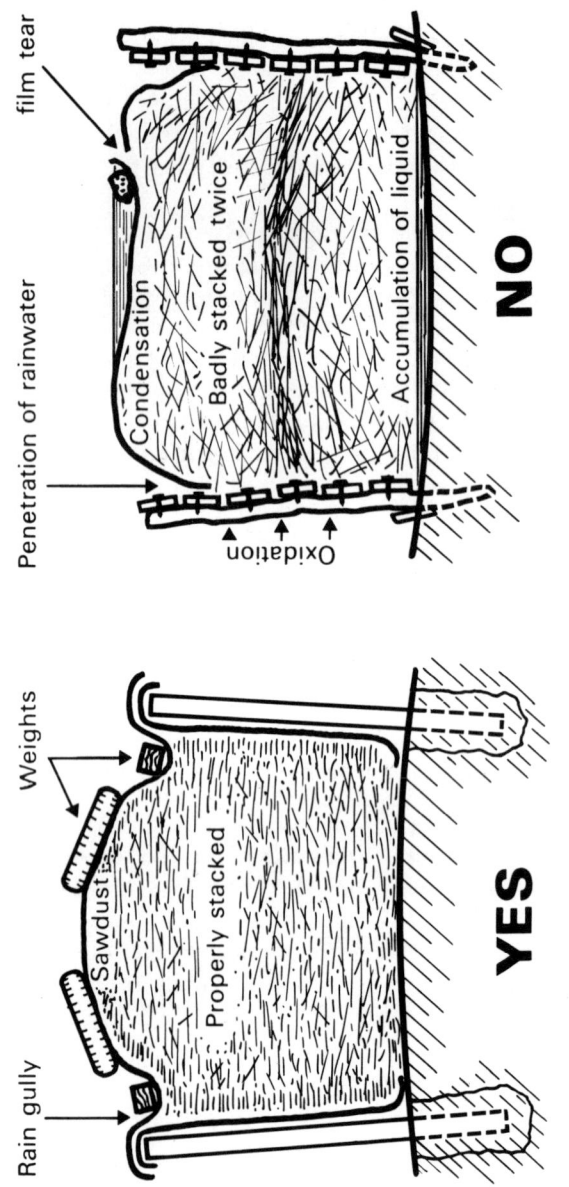

Fig. 16. Stacking of silage.

Note 1: Stack consolidation. Consolidation of the fodder can be carried out by repeatedly driving over it with a lorry or tractor with wide tyres, in the case of an ordinary clamp silo.

Note 2: Vacuum. This is a method which is generally applicable to grain towers and to silos having a liner of plastics film which is sufficiently tight.

Note 3: Carbon dioxide. Silage of maize under sheets of polyethylene enclosing an atmosphere of carbon dioxide.[70]

Equipment and Materials (See Chapter 5)
Silo Design

Capacity. Provision should be made to ensure that the capacity is sufficient to supply each head of cattle with 40 kg per day and, in the case of 'self-service', a feed trough of at least 20 cm width for each beast.

Base. The floor should be of concrete, and flat, with a slight slope towards the water drainage gully.

Reaping of the fodder. This is carried out in the afternoon when the fodder contains more sugar than in the morning; this provides a substrate for anaerobic fermentation enabling the production of lactic acid. The fodder is dried in the sun and then filled into the silo.

Installation of film. The stacked fodder is enveloped as completely as possible in a single operation (Fig. 16) both above and below with the sheets stuck together if possible.[71]

Air evacuation. The fodder must be compressed as much as possible during stacking (see also Chapter 5).

Loading. The top of the silo, particularly in the clamp type, is weighted down, either with soil or other weights, on top or at the edges to prevent the wind catching the film (Fig. 17).[72]

Note: Even at $-15\,°C$ beetroots do not freeze under a film of $30\,\mu m$ thickness, covered with a 30 cm thickness of straw which is itself covered with a black film of $100\,\mu m$ thickness.

Stalling or 'Self Service'[71]
Figure 18 shows a mobile shelter which is designed so that the animals can feed freely at any time.

Farm Buildings
Polyethylene sheeting can be used to provide relatively cheap farm buildings, particularly for animal protection. Black films containing 2% carbon black are favoured because of their high durability but a

FIG. 17. Silage in France ('clamp silo').[72]

disadvantage is the heat absorption during sunny periods; this can be overcome by painting the film with aluminium or white paint. A material approaching the ideal for covering single-skin buildings is a black polyethylene film to which is laminated a white reflective one.[74]

Where thermal insulation is required, a double-film structure, with the space between the two films being maintained by air pressure from a small fan, can be used.

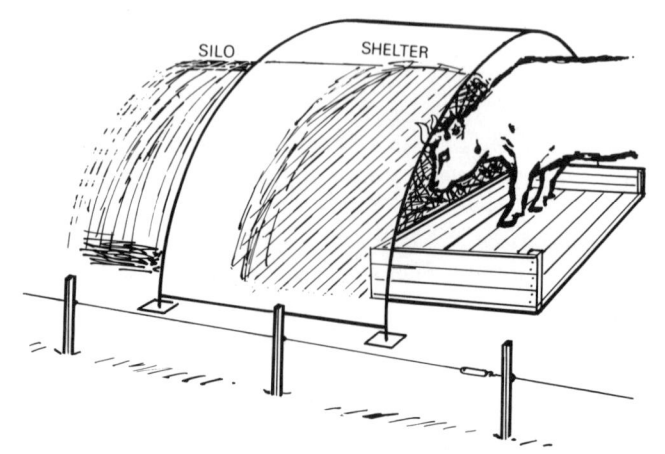

FIG. 18. Portable shelter for cattle feeding.

Steel tubing is an excellent structural material for the supporting framework but because of its expense, the tubing will need to be widely spaced with strands of high tensile wire or polypropylene twine between the steel members to give extra support to the plastics sheeting.

Mushroom houses are constructed to a similar design with a 4 in thick layer of glass wool to provide insulation between the inner and outer films of polyethylene (Fig. 19).

Water in Plasticulture

The map in Fig. 20 shows the magnitude of the rainfall deficit in France which, moreover, varies from year to year. Watering is therefore necessary, to some extent, everywhere. According to the French Ministry of Agriculture, the areas which are irrigated total 682 939 ha and these are indicated by Quentin.[73]

Role of films, tubing and reservoirs. They enable the large quantities of water which are necessary to be available to plants

(a) through storage by the creation of reservoirs and hill lakes, the construction of dykes and the control of streams;
(b) by regular distribution in artificial micro-climates (e.g. by mulching and in greenhouses); and
(c) by means of irrigation and also by the removal of water by drainage.[75]

Amount of water required by plants. The following list shows the amount of water in kilogrammes required to produce 1 kg of dry vegetable matter:

Maize	216	Corn	399
Clover	260	Hay	438
Peas	290	Lucerne	600
Potatoes	300	Beetroot	630

It is therefore necessary to supply to the plant the difference between the quantity of water necessary for its needs and that which it receives from the soil, except in the case when the latter requires drainage.

Synthesis requirements. Plants, like all other living things, make use of water for the synthesis of their tissues. This synthesis is brought about by the action of chlorophyll under the influence of radiation from the sun and also of environmental factors such as wind, which encourages transpiration and evaporation (together constituting evapotranspiration (ETP).

Evapotranspiration. If there is a shortage of water in the soil, then the plant closes its stomata. It therefore no longer breathes and this slows up

50   PLASTICS IN AGRICULTURE

FIG. 19. Mushroom house at Lee Valley Experimental Horticulture Station.

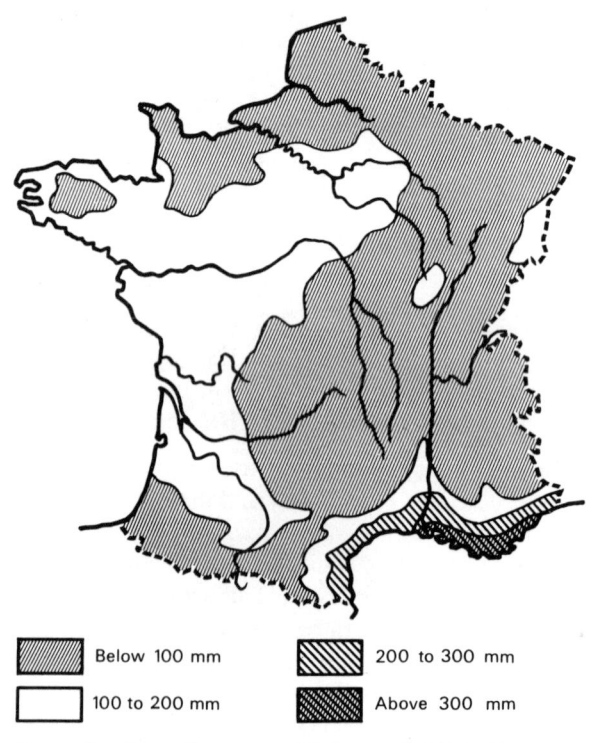

FIG. 20. Annual average deficit in rainfall in France.

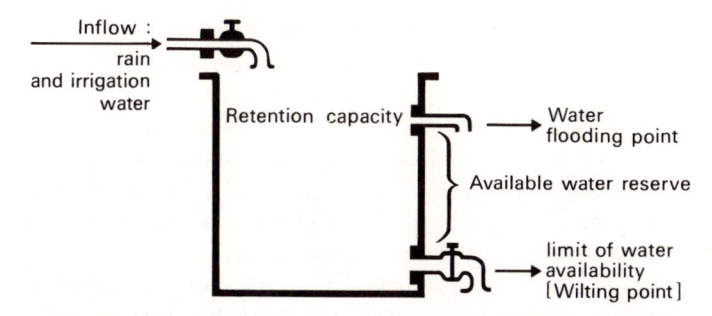

Fig. 21. Schematic representation of soil water retention capacity.

the photosynthesis which can also happen if there is a lack of radiation. The crop yield, which depends on the association of water, sunshine and fertiliser, diminishes.

The *utilisable water reserve* (UWR) available to the plant, as shown in the schematic diagram in Fig. 21, is the difference between the retention capacity of the soil and the limit of extraction; this capacity is itself dependent on the type of soil, as shown in Table 11. In arid zones however, in spite of the considerable suction power of the roots which can be up to 16 kgf/cm^2, the extent to which water can be removed from the soil is only about 30% of the available capacity; elsewhere this can be up to 50%.

TABLE 11
WATER RETENTION PROPERTIES OF DIFFERENT SOILS

Soil type	Retention capacity (%)	Capacity at wilting point (%)	UWR (%)
Clay	35	18	17
Silt	18	9	9
Sandy silt	13	6	7
Sand	6	2	4

It is clear that the water which is readily available to the plant (the UWR) is only a fraction of the water received by the soil because the latter finishes up in different places, i.e. as run-off water, seepage water lost or diverted, and water which is a constituent part of the soil and is not extracted by the roots.

The limit of extraction of the roots is such that if it is insufficient, the plant wilts: this is termed the permanent wilting point.

Water requirements. Each plant requires a certain volume of soil for occupation by its roots, for example up to 50 cm in depth for orchards. The water requirements are variable depending on the district and the season (Table 12) and must be satisfied for each period of growth; month by month a quantity of water is supplied to the plant and a quantity is lost by evapotranspiration. Thus, from day to day, the deficit can be made up if necessary, since 70% of the retention capacity is found below a depth of 30 cm. These water contributions are controlled in terms of the speed of absorption or of the toxicity (residual water); spraying seems to be the quickest and most effective method of supplying the water but trickle (drip) irrigation has given a new aspect to the problem.

TABLE 12
POTENTIAL EVAPOTRANSPIRATION IN VARIOUS REGIONS

Area	Spring			Summer		
		Monthly average (mm)				
France	April	May	June	July	Aug.	Sept.
Northern (Lille)	56	79	77	79	77	57
Central (Orleans)	100	113	121	135	135	93
South-west (Toulouse)	69	89	110	135	136	105
Mediterranean area	120	184	253	266	207	170
Copenhagen (Denmark)	Av., Nov.–April = 11			Av., March–Oct. = 78		
Valentia (Ireland)	Av., Oct.–March = 17			Av., April–Aug. = 66		
Tunis (Tunisia)	Av., Oct.–March = 55			Av., April–Sept. = 170		
Dujailah (Iraq)	Av., Jan.–April = 85			Av., July–Sept. = 248		

Soil Sterilisation

A note on crop rotation. This is the methodical succession of crops so as to obtain maximum yields from the soil without impoverishment; it can be every 4, 3 or 2 years. In the first, the English Norfolk Rotation follows the sequence: first year, beetroot or potatoes; second year, summer cereals; third year, clover or grass; fourth year, winter wheat. The triennial rotation adopted in Brie and Beauce in France follows the sequence: first year, sugar beet; second year, wheat; third year, oats. The growing of leguminous vegetables after wheat restores the nitrogen to the soil which has been taken up by the latter.

The object of fumigation. Soil sterilisation allows the same crop to be grown year after year on the same plot and it also cleanses the ground in getting rid of weeds and of pests such as nematodes, insect larvae and micro-organisms responsible for plant diseases.

Principles of sterilisation. Formerly it consisted of injecting into the soil, to a depth of several centimetres, steam at 80 °C, or volatile chemical products, using perforated tubes of about 3 cm diameter and spaced 200 cm apart, while the area is covered with a sheet buried at the edges. The sheet inflates under the pressure. In 10–20 min, units of 6 m × 10 m in area by 10 cm in depth can be treated.

Fumigation is based on injecting into a well-prepared plot, covered with polyethylene film stuck down at the edges, a toxic liquid which volatilises only slowly so that the vapour is maintained in contact for sufficient time to destroy all the plant and animal parasites.

The principal *fumigants* in use are *chloropicrin* (trichloronitromethane) and *methyl bromide* which act synergistically; they are mixed in various proportions according to the destructive effect required, the first material is used to control *Verticillium alboatrum* in strawberries and the second for the destruction of weeds. Without the covering of polyethylene it is necessary to use 216 kg of fumigant for 40·5 ares (1 are = 100 m^2) for strawberries and this is reduced to 140 kg with a cover. Other fumigants can also be used for nematodes and a covering is not always required.

For the *preparation* of the soil all roots should be eliminated and watered for several months in advance, in order to speed up their decay. If necessary, clods should be broken down mechanically using a disc harrow, after watering should this be required. The best temperature for this work is between 10 and 24 °C. It is finished off by levelling the soil.

The *equipment* used in soil sterilisation is described in Chapter 5 (see Fig. 58).

Fumigation allows crops of high profitability to be grown without interruption on the same ground from year to year; the crops include strawberries, tomatoes and flowers.[76]

The cost of the first treatment is relatively low in comparison with the costs of hoeing which would be necessary. The subsequent treatments are also less onerous. This method can be applied to different types of ground, from sand to clay.

PLASTICS SHEETS, LINERS AND VESSELS

General

In addition to the films used on a large scale for the protection and development of plants, e.g. mulching tunnels, greenhouses, silage, packaging, etc. an increasing quantity of flat and corrugated sheet is employed for construction purposes in farm buildings, stores, fences, containers (e.g. tanks, vats, reservoirs and silos) and agricultural machinery (e.g. threshing and reaping equipment, etc.).

Properties
The properties of particular concern are not only the mechanical strength, ageing resistance and thermal insulation but also the electrical conductivity which can lead to the development of *electrostatic charges* by friction with liquids or solids.

Principal Materials
The principal materials in use are resins and polymers, notably PVC, polyesters, styrene polymers (ABS) and polyethylenes including EVA, polypropylene and phenolics. These are reinforced by means of various fillers, e.g. glass fibres, slate and mica. Examples of their applications include mushroom houses in glass reinforced polyester insulated with polyurethane foam;[77] load-bearing structures in glass-reinforced polyester;[78, 79] plastics coatings for machine components by dipping;[80] anticorrosion of metal parts in greenhouses;[81] and anticorrosion tank linings.[82]

POWDERS AND GRANULES

Soil Aerators
The lightening of heavy soil can be effected by the addition of foamed plastics in granular or chip form. The initial work was carried out with expanded polystyrene in order to establish a use for what was at the time a waste product.

This material is now established as a soil additive which improves soil structure and stimulates root formation so that it finds increasing use for plant propagation and for pot plants. It is also used for drainage in place of conventional drains. Urea formaldehyde foam is also used for this application. It has the advantage that it is moisture retentive (unlike polystyrene) and decomposes slowly in the soil to supply nitrogen to the plants.[83-85]

TUBING AND PIPING

General
As in other fields of application plastics compete with numerous traditional materials with considerable success. Both low and high density

polyethylene and rigid PVC pipes are used for the transport of fluids (both air and liquids), they are unbreakable and do not corrode and are replacing various alloys and rubber because of their light weight and ease of installation.

In plasticulture, piping is used particularly for potable and waste water, for irrigation and drainage and also for beverages (milk, wine, beer, cider) and various solutions (fertilisers, pesticides, etc.).

Irrigation
This aims to satisfy the enormous water demands of plants to ensure their proper growth.

Channel Irrigation
This is the system which has been used since ancient times and is still in operation in many countries. Water is fed from a main channel into a series of ditches which are created when the plants are grown in ridges[86] (Fig. 22a). The flow of water from the main channel into the gullies can be

FIG. 22a. Channel irrigation. (By courtesy of J. Hanras.)

FIG. 22b. Channel irrigation with polyethylene siphon tubes. (By courtesy of J.-C. Garnaud of the CIPA.)

FIG. 22c. Glass-reinforced plastic conduit and siphon. (By courtesy of Scott Bader Co. Ltd, Wellingborough, Northants., UK.)

controlled by using polyethylene siphon tubes (Fig. 22b). An up-to-date version of this technique, using gullies made from glass fibre reinforced polyester is being used in Malaysia for irrigating rice fields (Fig. 22c); these have replaced concrete channels. A simple S-shaped siphon is hooked over the end of the conduit at any place along its length, making it simple to dispense the water where it is needed.

Spray Irrigation
Water is taken from a river, well, lake or reservoir and is pumped through a distribution system to feed sprayers or sprinklers which are spaced at regular intervals over the ground to be watered.

Note 1: The power requirement P of the pump is given by the expression:

$$P = \frac{\text{Delivery (litres/sec)} \times \text{Delivery pressure}}{75 \times \text{Coefficient of efficiency (50–65\%)}}$$

About 0·1 kwh or 0·035 litres of gas oil or 0·05 litres of petrol is required to deliver 1 m^3 of water at a height of 25 m. The service pressure is generally low (1·5–5 kg/cm^2 (about 20–65 psi)).

Note 2: The distribution network is generally as shown in Fig. 23.

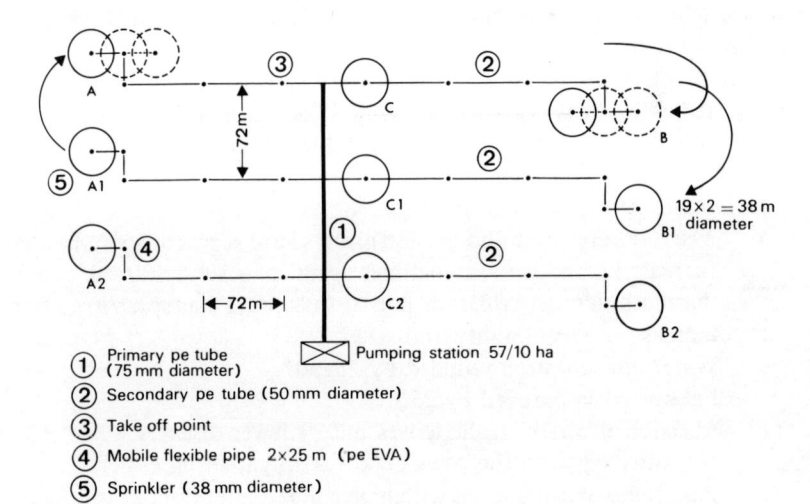

① Primary pe tube (75 mm diameter)
② Secondary pe tube (50 mm diameter)
③ Take off point
④ Mobile flexible pipe 2×25 m (pe EVA)
⑤ Sprinkler (38 mm diameter)

FIG. 23. Irrigation pipe network, pe = polyethylene.

Note 3: *Pipe performance* is of particular importance for the installation and operation of the distribution system (see 'Pipe Testing' (Chapter 6)).

Note 4: *Anti-frost measures.* Spraying is often used since it is less onerous than heating; local freezing of water requires the removal of 80 kcal/litre which leads to a local reheating.

Trickle or Drip Irrigation

The system of trickle or drip irrigation was originally developed in the Negev Desert in Israel because irrigation by spraying the sand with salt water gave unsatisfactory results. It is now used in many countries and a whole range of different commercial systems is available. It is used especially for strawberry cultivation in California. The technique is gaining popularity in those areas where water supplies are inadequate. Many workers have engaged in the evaluation of the different systems, particularly in relation to the crop under cultivation.[87-94]

Principle. The distribution pressure of the water is reduced from 15 to 25 psi to a few feet head at the point where it is fed to the roots of the plant at a very low rate as a trickle or series of drips. There are many systems by which this can be accomplished: by use of a porous plastics tube, by a micro-tube from a main feeder, by a perforated tube, or by means of dripper devices fixed to a feed pipe. The quantity of water can be controlled so that optimum results are obtainable with any particular crop under different climatic conditions. It is essential to use filtered water to avoid blockage of the feed system.

Areas of applications. For crops grown at wide spacings such as orchards tubes fitted with drippers are used. Layflat hose systems are used for crops grown fairly closely together such as vegetables in the open, and flowers in greenhouses.

Advantages of trickle or drip irrigation

The roots are never flooded and always have access to air and water.
Increase in crop yield, sometimes as much as 100%, 20–40% more than with channel irrigation, and 10–20% more than spray irrigation.
Increase of crop quality and size.
Water consumption reduced by 25–50%.
Labour costs reduced by 25%.
Absence of water on the leaves means fewer diseases.
No water between the rows makes crop gathering easier.
Possibility of using water with high solids content since the salts are not deposited at root level.

Disadvantages
    Cost is higher than in other systems.
    Relatively short life (particularly with layflat tubes).
    Blockages of perforations, holes and pores may occur.
    Attacks by rodents and animals seeking water in dry regions (e.g. Arizona in the USA, South Africa).
    Installation: all the layflat irrigation tubes must be run out over the soil with a tractor and sometimes buried to a depth of 1–4 in to avoid attack by rodents and blockage of the jets by evaporation of the water drop at the tip.

Use of mulch with trickle irrigation. This technique is particularly effective for making the best use of the water which is being fed to the plants. It is now used extensively in the USA for the cultivation of strawberries; apparently, surface application, with water constantly moving downward, improves the plant performance and this is only possible when a mulch of polyethylene film is used. This system is also capable of being used with a dilute solution of fertiliser.

Significant improvements in the yields are obtained, particularly with low-growing plants such as melons. The gain in the earliness of the crop is not so pronounced with taller-growing plants such as tomatoes and peppers.

Irrigation in Greenhouses

The amount of water required is 600 to 1000 mm per year, according to the nature of the soil, the stage of plant growth and the temperature, in three operations:

    Water reserve. This is quite small and of the order of only 5 mm; it maintains a humidity favourable to gaseous exchange with the liquids and solids in the soil.

    Watering. From 20 to 40 mm of water is required for the diffusion in the soil of the nutritive elements needed for plant growth.

    Soil washing. Twice a year, the soil is washed with 200 mm of water in order to eliminate the complex salts which come to the surface by osmosis (e.g. sulphate and chloride of ammonia). After being dissolved the salts are removed through the drainage system and the control is made at the collector by the analysis of the water and the specific gravity.

    Mist spraying. This involves supplying small amounts of water as a fine spray in order to maintain a constant humidity while avoiding scorching and the encouragement of microbial growth and at the same time economising on the amount of water used.

Dormancy. Plants become dormant when the soil and the atmosphere are allowed to become dry.

Drainage
Large areas are periodically flooded in France and this sometimes means the abandonment of large tracts of good land. It is reckoned that there are 6×10^6 hectares (15×10^6 acres) which could be drained. In the UK about $1 \cdot 25 \times 10^6$ acres are drained annually, using approximately 150×10^6 ft of pipe.

Soil permeability. This depends both on the soil structure and on its composition. Sand is made up of particles with an average diameter of 0·05 mm, silt of 0·002 to 0·02 mm diameter particles, while clay and humus have an average particle size of 0·002 mm (2 μm). The finer particles can plug up the interstices between the larger ones, thereby preventing the flow of water through the soil. Under such conditions the roots can become asphyxiated and water will collect on the surface.

The rate at which water seeps away depends not only on the type of soil but also on the depth at which it is determined. This is expressed as the amount of water in a given time per square metre of surface which seeps away from the sides of open trenches cut for laying the drains. This is shown in Fig. 24 for various types of soils; if the flow is less than 0·005 litre/sec/m² of trench wall surface then drainage is required.

Surface drainage. It is sufficient sometimes to provide drainage channels by cutting through the impermeable layer and running off the rain-water from the shallows where it collects.

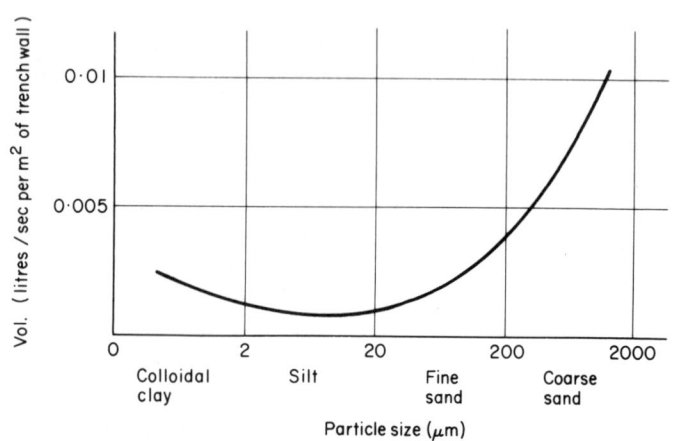

FIG. 24. Water drainage from trench walls.

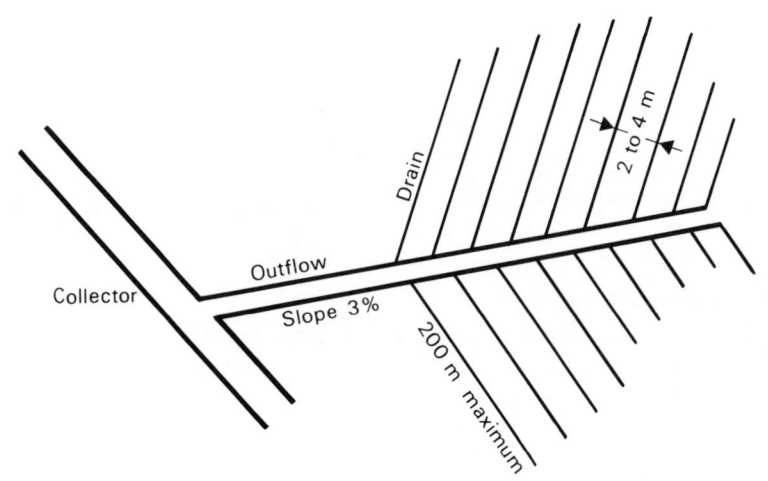

FIG. 25a. Drainage network.

Deep drainage. It is often the case that the topsoil lies on an impermeable compact layer with practically no slope. Under such conditions a drainage system is required.

There has been a rapid development in the use of plastics pipes for the replacement of drainage tiles, because of their light weight, flexibility and corrosion resistance. In Europe, rigid PVC is normally used, and there is some interest in high density polyethylene; in the USA, the latter material has dominated the market.

The design of the system required must be based on the topography of the subsoil, the filtration speed, the water table and the impermeable zone, and it must include the drainage spacings, the length of drains and their location in relation to the outflow and the collector (Fig. 25a). This can be done by

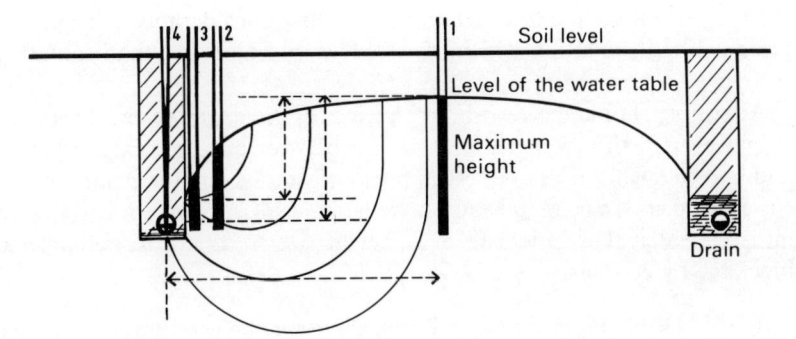

FIG. 25b. Level of the water table between the drains.

the use of piezometers to define the spacing of the drains which are in the form of tubes with longitudinal slits (Fig. 25b).

Note: The major developments in the use of plastics pipe for drainage have come from The Netherlands, where about 12 000 km are in service, and where work continues in this application with studies on the frequency of drains, and on the form of the slits and their position in relation to periphery of the tube.

*Drainage and Irrigation**
Irrigation with distribution systems. These are based on the distribution of water, generally by means of a network of pipes using a variety of filters for the removal of solid matter. Differentiation should be made between the methods and conditions of supply (pressure and means of conveyance) and also the method of water distribution in the soil. Spray irrigation does not necessarily involve the transport of the water by pressure through a water-main for a small plot, since this can be readily watered. In this case, only a single pump is used, although an automatic pumping station containing several pumps is required for a system covering large areas of ground. In this case, the formula for pump power does not generally apply.

The connection of irrigation systems to the main supply of potable water is not recommended since the cost for such water is much too high. The pipework system is not designed for irrigation requirements subject to pressure surges due to the opening and closing of the valves. The overloading and reduction of pressure appears to accelerate the ageing and deterioration of plastics pipes. It is therefore not considered advisable to connect irrigation systems directly onto a communal supply.

As regards drip irrigation, the main obstacle to the greater use of this technique lies in the difficulty in removing the suspended colloids from the water with anti-colloid filters. The price of these considerably increases the price of the installation which, by itself, would be lower than for other systems.

Attention should be drawn to the use of rigid perforated pipes for irrigation, a method which is intermediate between drip or trickle irrigation and the classical techniques. It appears advantageous for plants grown close together. It can be fully automated and fertilisers can be incorporated into the water. The holes are of sufficient size so as not to be liable to blockage by colloids.

* By J. L. Dervil, Ingenieur Genie Rural, Direction du Genie Rural, 19 Avenue du Maine, Paris, France.

It should be mentioned also that it is essential to avoid concentrations of harmful substances (e.g. pesticides, heavy metals, etc.) when drainage water is collected for irrigation.

Finally, drip or trickle irrigation does not cause leaching or compaction of the soil, which can occur with other methods.

Drainage. It is no longer considered necessary to bring under cultivation all available, exploitable land, for the reasons of water supply, ecology and economy. It is necessary to retain fens and marshland since they maintain the amounts withdrawn from the rivers which supply water to large numbers of animals and plants.

Mechanisation appears to be leading more and more to an end of the use of clay tiles and pipes in the agricultural sphere. This requires a strict control of the depths of the excavation, which becomes more and more difficult with machines of higher and higher speed in spite of the use of lasers as a method of control. Clay tiles and pipes are heavy and more expensive, and plastics pipe delivered as a coil allows installation on reverse slopes. It does not become disconnected, even with settlement of the ground (which occurs particularly in peat-bogs).

MISCELLANEOUS APPLICATIONS

Equipment and Machinery

There is an increasing use of plastics for components such as covers and bearings in agricultural and horticultural equipment. Both polypropylene and polyamide (nylon) are materials which can be moulded to give high strength components. Extruded sheet can also be vacuum-formed to produce covers and boxes.

Growing Trays and Troughs

Extensive use is now made of trays and troughs moulded from polypropylene and high density polyethylene. A typical development is the 'Empot' handling system, in which potted seedlings are grown in trays carried five high in brackets on two sides of a free-standing pillar in the greenhouse. The various components are made from polypropylene and can be easily handled onto a trolley for transporting.

Double wall polypropylene extruded sheet (Correx) is light and rigid and capable of being readily cut and fabricated. It is used for forming the gullies lined with black polyethylene film in hydroponic cultivation (see Chapter 5). A similar application is for the canal system now being used for growing tomatoes in Jersey. Continuous troughs in 1–2 m lengths are formed from black sheet and are filled with a peat compost which need not be renewed for at least 5 years and may be sterilised in position before each growing season.

Produce Handling and Packaging

A wide range of trays and crates, moulded from polypropylene and high density polyethylene is now available. A new range of produce boxes, the 'Euro' stacking range, based on a plan size of 600 mm × 400 mm with differing depths between 75 and 320 mm is now being marketed; these are completely interstackable.

A thin polypropylene sheet with folding qualities is being produced as a replacement for carton board in packaging. Strong, ventilated cartons may be produced economically for handling produce such as watercress; the material which is cut out for making the ventilation holes can be re-used by the manufacturer to produce further cartons. Boxes produced from this material are suitable for cut flowers since they provide good protection for the blooms.

Plastics for produce handling have been the subject of much development work in Italy, where a whole range of crates and boxes is produced in several materials. One of these uses structural foam high density polyethylene, reinforced with 10% glass fibre. It is claimed that very good strength is obtained and crates for harvesting and transport of fruit such as apples are performing extremely well.

Fig. 26. Strip polyethylene film for shading.

Shading

This can be done using either coloured films or polyethylene mesh screens on either the interior or the exterior of greenhouses. Black polyethylene film mounted in the form of an easily movable tunnel is used to control the day length in the cultivation of chrysanthemums out of season.

In those parts of the world where shading is needed to protect an area from excessive sunshine, several techniques, mainly using film, have been described.[95] A low-cost and easily erected shade area can be made by tying pieces of black polyethylene sheeting to strings or wires stretched above the crop (Fig. 26).

FIG. 27a. 'Plantube' propagation pot.

Containers and Pots

Plant pots moulded in polystyrene or polypropylene have almost completely replaced the clay pot both for commercial undertakings and for the private gardener. The advantages are well-known: the pots are easy to clean, of lighter weight and evaporation losses through the sides are considerably reduced.

Black polyethylene film containers are cheaper and consequently have become accepted as the norm by the majority of nurserymen. Small containers from polyethylene mesh are also used.

A special design of cylindrical pot (Fig. 27a) of varying sizes, formed in polystyrene is widely used in France (the Plantube). Its height is several

FIG. 27b. Use of 'Plantubes' for growing cacti.

times its diameter and the sides are formed with helical shaped grooves which encourage the formation of roots through the whole depth of the pot. This type of container is widely used commercially (Fig. 27b) and has been used for the export of young vine plants to Nicaragua for the establishment of vineyards. They have also been used to transport young eucalyptus and pine plants for the creation of a green belt in Algeria over a width of 20 km between Tunisia and Morocco, in order to modify the climate and to reclaim some of the desert areas.

FIG. 27c. Cultivation of mushrooms using polyethylene sack containers.

With the construction of more and more multi-storey housing there is an increasing demand for plant troughs which are designed so that there is a constant supply of moisture available to the plant roots. Such containers can be readily moulded using polypropylene.

Polyethylene film sacks are used for the cultivation of mushrooms in underground quarries (Fig. 27c).[96]

Tanks and Vessels
There is an increasing usage of tanks in plastics materials for handling corrosive aqueous solutions such as fertilisers and herbicides. These are generally produced from glass-reinforced polyesters (GRP), although the technique of blow-moulding thermoplastics has enabled larger sized vessels to become available, using polypropylene and high density polyethylene.

Vessels in GRP from 1000 to 40 000 gal capacity have been in use for many years for the storage and processing of fruit pulps, fruit juice concentrates, wine, edible oils, and aqueous solutions on the farm.

Baler Twine
This particular application has become dominated in recent years by polypropylene which is oriented to give a high tensile strength. The ageing has been improved by incorporation of UV stabilisers which are effective for several seasons.

Protective Sleeves
Thin gauge pigmented blue polyethylene film sleeves as a loose covering on banana bunches during the growing season were first introduced by the New South Wales Department of Agriculture in the early 1950s. The colour has no specific technical advantage and is intended as an identification of the material. Polyethylene is preferred to flexible PVC because of its lower price and the covers are used for only one season.[97]

FIG. 28. 'Lammac' polyethylene protective coat.

Animal Protection
During the early part of the year losses amongst lambs because of exposure to a combination of wind and rain can be high and 'Lammacs' have been introduced as a protection for the young animals. The small coats (Fig. 28) are made from polyethylene film in two sizes and two colours, one for singles and one for twins.

Chapter 4

Properties of Semi-finished Products in Plasticulture

COEFFICIENT OF EXPANSION

In all assemblies containing different types of materials, it is essential to take account of the differences in the coefficient of expansion, which can be considerable as is evident from the following orders of magnitude:

Silicon, silicates and aluminium silicates	$0-10 \times 10^{-6}$
Metals and alloys	$0-30 \times 10^{-6}$
Resins and polymers	$0-300 \times 10^{-6}$

It is therefore clear that certain resins and polymers can on average expand by a factor 30 times more than metals and alloys. This leads to the development of high stresses in assemblies containing different materials.

The expansion of resins and polymers can be modified by the incorporation of mineral or metal powders as fillers and also by use of a plasticising glue;[98] this latter gives to the joint the possibilities of deformation without rupture.

MOLECULAR ORIENTATION AND ANISOTROPY

When ductile thermoplastics polymers are stretched, the long molecules become less entangled and become aligned so that the material becomes thinner but increases in strength (Fig. 29a). In the case of a fibre filament the orientation (OR) is expressed by the ratio:

$$OR = \left(\frac{\text{initial diameter}}{\text{final diameter}}\right)^2$$

This is the *degree of stretching* illustrated in Fig. 29b, which shows that

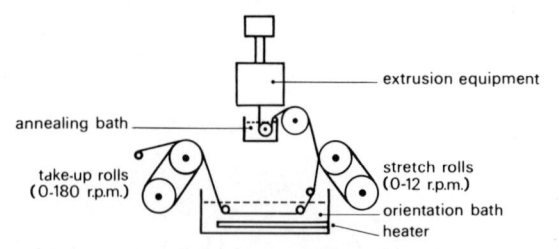

FIG. 29a. Orientation of monofilaments by stretching, cooling and re-heating.

the rupture strength increases and the elongation at break decreases as the *OR* ratio increases.

In the *extrusion blowing* of films (Fig. 29c) the tube which emerges from the die is thick, with a diameter of 10 cm; this is then increased in size by inflation with air.

This has the effect of stretching the film in the transverse direction so that there is some disentanglement and orientation of the molecules; the film take-off equipment also creates some stretching in the longitudinal direction so that the final film, after a considerable increase in diameter, has a measure of orientation in two directions at right angles, giving a material with, for example, a thickness of 50–100 μm.

If the orientation is equal in the two directions at right angles, the rupture strength in the two directions will be the same and so also will be the elongation at break if the two samples are taken in the longitudinal and

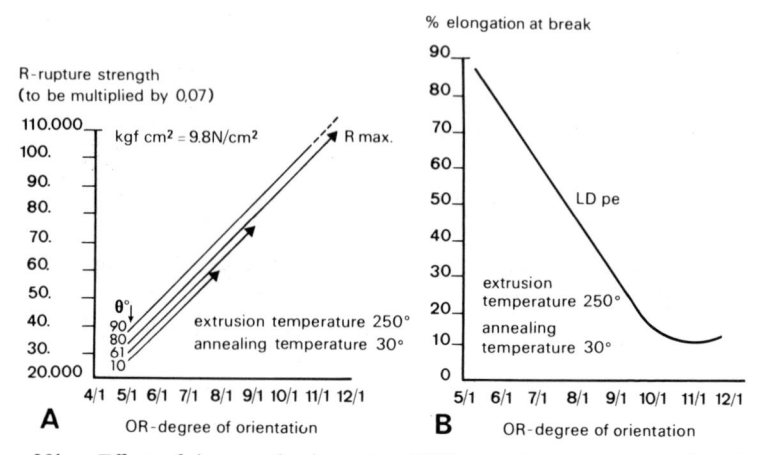

FIG. 29b. Effect of degree of orientation (OR) on (A) rupture strength and (B) elongation at break.

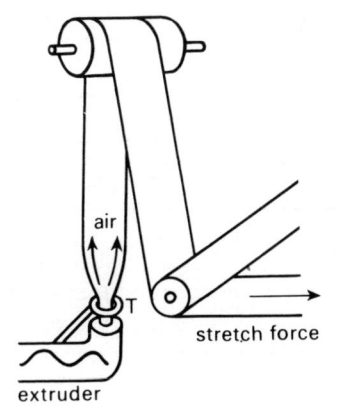

FIG. 29c. Extrusion blowing of film.

transverse directions. In such a case, the film has anisotropic properties, and is referred to as a 'balanced film'.

If, however, a film is produced on a calender the rupture strength in the running direction is much higher than in the transverse direction across the film.

When transverse stretching is also applied to a calendered film then some measure of anisotropy is introduced and a bioriented film is produced (as in the case of polystyrene, for example).

If a film is produced by evaporation of a solution of the polymer in a solvent there is no orientation because the molecules are deposited in a random fashion. Figure 29d shows the effect on the rupture strength and elongation at break of stretching 'Pliofilm', a chlorinated rubber; some measure of orientation is introduced at relatively low tension in a system where the molecules are entangled in a random fashion.

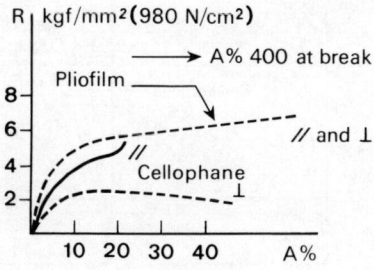

FIG. 29d. Rupture strength and elongation after stretching (Pliofilm and Cellophane).

PLASTICS IN AGRICULTURE

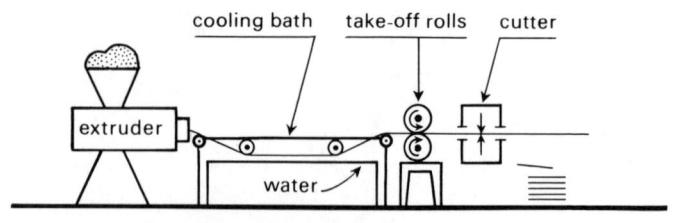

FIG. 29e. Sheet extrusion.

Note 1. *Polypropylene tapes* are produced by differential stretching of the sheet, which becomes highly oriented, and this is then cut by circular saws into tapes of varying width according to the application (e.g. 3 mm).

Note 2 *Extrusion of pipes, rods and sheets.* The extrudate, after emergence from the die, is passed through a sizing die, in the case of pipes and rods, after which it is cooled by passage through a water-bath. In the case of sheet extrusion, the extruded material passes through nip rolls and is then air-cooled. The products are cut to length after having cooled and become rigid (Fig, 29e).

PROPERTIES UNDER TENSIONAL STRESS

Depending on the material, the temperature and the rate of extension, the curve for the stress plotted against the extension can take many forms.

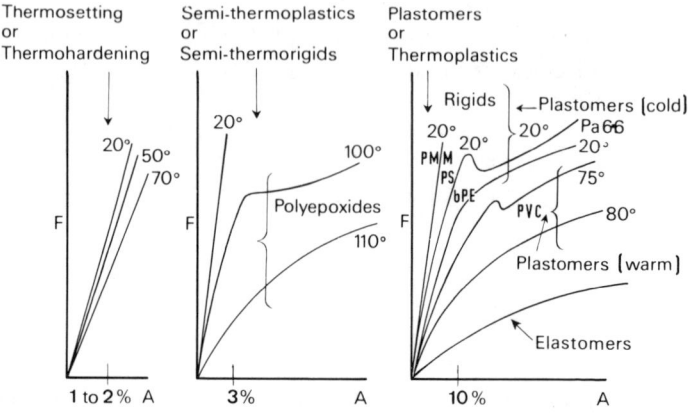

FIG. 30a. Stress–strain diagram for different types of plastics materials.

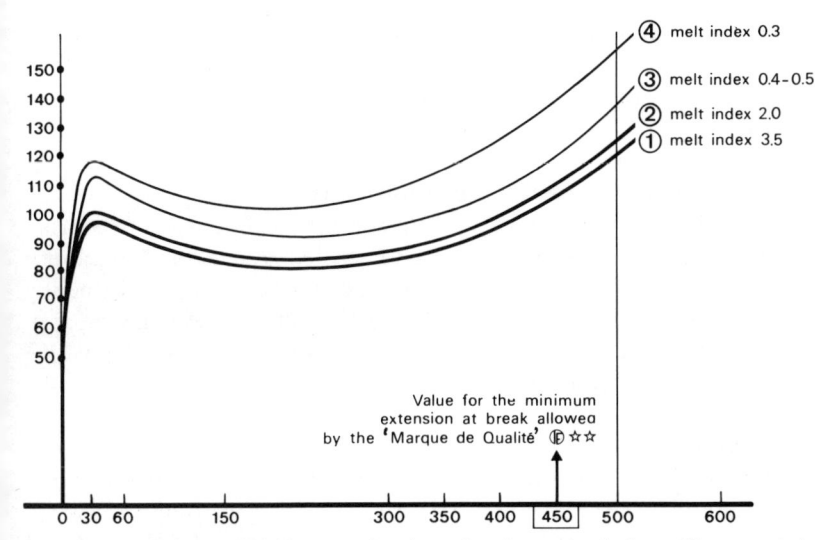

FIG. 30b. Stress–strain diagram for low density polyethylene films used in agriculture. Ordinate: stress (kg/cm^2). Abscissa: extension (%).

Examples are given in Fig. 30a for the thermoplastics (PVC, low density polyethylene, polystyrene, Nylon 6.6 and polymethyl methacrylate) and for the thermorigid materials (polyesters, glass fibres and phenolics) at the same rate of extension in the different cases.[99] The change in elongation at break can be measured to indicate any deterioration or ageing of the materials, although this is quite low for thermorigid materials.

When a stress is applied to thermoplastics there is a level at which the material yields, i.e. extends, without further increase in the applied force, and this is clearly shown on the curve for PVC at 75 °C and on that for polyamide 6.6 (Pa 6.6) at 20 °C. In the case of low density polyethylene, a tangent is drawn at the origin of the curve and then a parallel line at a conventional displacement of 20% which cuts the curve at a point and arbitrarily determines the yield point. The variation of the elongation at break in relation to the stress and the grade for polyethylene is shown in Fig. 30b.

AGEING AND THE QUALITY STANDARD

Factors Affecting Ageing

Plastics materials degrade by photo-oxidation with the partial breaking of the long molecules into shorter components; this arises mainly by the action of UV radiation although water does play a part, particularly in the case of polyesters where some measure of hydrolysis occurs. The degradation is reflected in a tendency for the plastics to embrittle and the elongation at break decreases.

Correlation Between Material and Artificial Ageing

In view of the variable nature of natural weather conditions, it can be risky to rely on artificial ageing experiments carried out in the laboratory, because of the difficulty of precise duplication of natural weathering conditions. However, by taking an average of results over several years, e.g. 5 years, one has the chance of characterising a climate as nearly as possible from the point of view of the fluctuations of temperature according to the season and the year.

Macskasy and Szabo[100] have made comparative studies on the change in the mechanical properties of low density polyethylene during the course of natural and artificial ageing (Fig. 31). Laboratory ageing, using a xenon arc lamp, is accelerated considerably so that the elongation at break is reduced by 20 % in about 400 h as against more than 2000 h for natural ageing to the

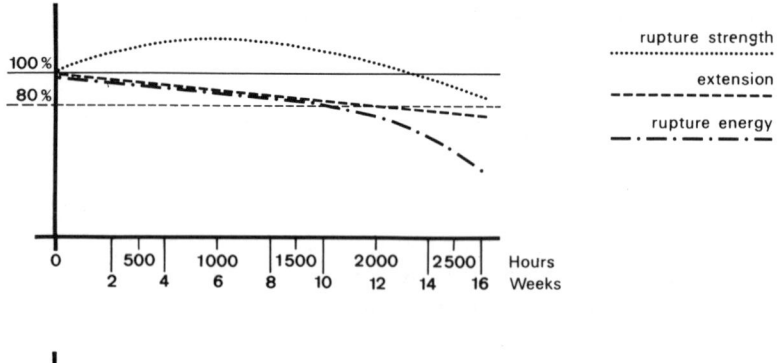

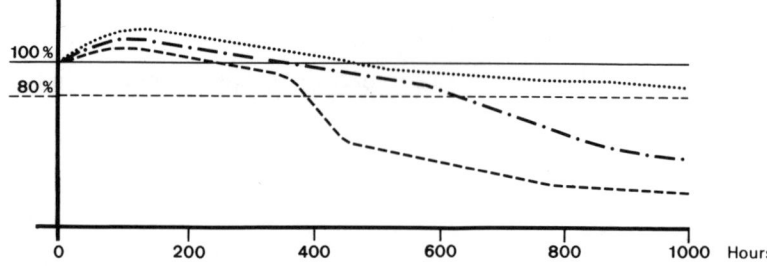

FIG. 31. Change of mechanical properties of low density polyethylene on ageing; top: natural ageing in Budapest; bottom: artificial ageing (Xenotest).

same extent at Budapest. If the location of the use for the plastics is always the same, i.e. the natural ageing conditions are roughly similar from year to year, then some measure of correlation can be obtained between the relative times for natural and artificial ageing in order to arrive at a reduction of 50% in the elongation at break, which is the value considered to be the limit for the useful life of the material.

The assessment of the performance of polyethylene films by accelerated ageing in the laboratory is included in standards which have been published both by the Centre d'Étude des Matières Plastiques and the British Agricultural and Horticultural Plastics Association.[101,102] The methods in the two standards differ to some extent but the field performances of the films are equivalent. The CEMP standard designates two requirements, 'one-star' is considered suitable for at least a year in actual use and 'two-star' will last for at least 2 years. The BAHPA standard covers only film of 'two-season' quality, i.e. it shall perform satisfactorily as a greenhouse covering for two full summer seasons under UK climate conditions when used to clad tunnel or multi-span greenhouses based on acceptable design and correctly installed. The CEMP laboratory ageing test requires that the films shall retain at least 50% of the elongation at break after exposure for 8 days for one-star film and 13 days for the two-star film. The BAHPA test sets the same criteria for the equivalent of two-star, i.e. a 50% retention of elongation, but after 1000 h in a Weather-o-meter.

CONTACT EMBRITTLEMENT OF FILM

During the course of processing plastics materials, heating is followed by cooling to harden the material; because of the poor thermal conductivity of plastics the surface freezes while the inside may still be soft so that internal stresses can be set up during cooling. These stresses can then result in cracks in the material under some circumstances when it is put into use.[10]

Embrittlement can occur under the influences of mechanical constraint, radiation (IR associated with UV), certain chemicals (stress cracking agents), and electrical forces. The effect can be enhanced if there is an association of constraints and this is the case at the points where the film is in contact with the hoop supports (stress and solar radiation) in tunnel greenhouses.[103] Experience has shown that localised rupture where the film is in contact with the metal support can be reduced if the film is shaded from the sun over this area to prevent excessive heating. This can be done by applying to the outside a paint which is heavily loaded with titanium dioxide or aluminium oxide to absorb solar radiation. Vinyl or acrylic paints are suitable so long as they contain no substances which can cause stress-cracking.

Protection can also be obtained with a band of white-pigmented polyethylene tubular film containing titanium dioxide which is held in place over the line of the hoops by two cables which pass through the interior of the tube.

CUTTING OF FILMS AND SHEETS

Films are cut by sawing of the roll perpendicular to the length and then cutting if need be. Holes are produced with an electric drill (Fig. 32) or a perforator with a heated ring.

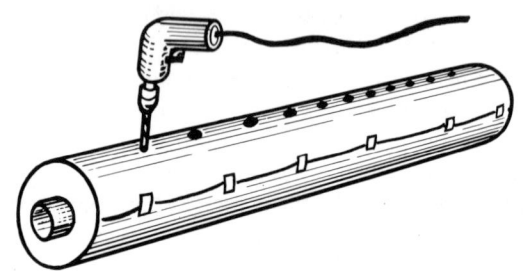

Fig. 32. Film perforating with electric drill.

Sheets (PVC, polyethylene and polyester) up to 1·5 mm in thickness are cut with clippers, up to 3 mm in thickness using a guillotine, and then at greater thicknesses with a saw such as is also used for tubes.

SHAPING OF SHEETS

For shaping, sheets are heated with a low flame on one side and then on the other to render them pliable but this is not possible with sheets of glass-reinforced polyester. Holes are made with a high speed drill with a tip angle of 130–140° at a rate of about 700 mm/min. Riveting can be done with polyamide rivets (Fig. 33).

Fig. 33. Plastics (polyamide) rivets.

ADHESION AND GLUING

Homogeneous Adhesion
This method of gluing a plastics material to itself is in principle easy, if a solvent for the material is available, but it cannot be done with thermosetting plastics (phenolics, amino plastics and polyesters).

Soluble Adhesives
A typical case is that of a solution of unvulcanised rubber in benzene; a PVC glue can be made by dissolution in cyclohexanone.

Note: A monomeric glue is possible by use of methyl methacrylate polymer (PMMA). This is used as a powder which is compounded with the monomer and a catalyst, for mounting artificial teeth.

Adhesive Emulsion
This is in the form of an emulsion of an ultra-fine powder in a solvent or water: aqueous emulsions of polyvinyl acetate are used as adhesives and as emulsion paints for buildings.

Note: The adhesive is often loaded with solid fillers such as wood flour, or with pigments in the case of paints, which increase the resistance to shear in the first instance and generally decrease it in the second.

Heterogeneous Adhesion
Different materials are stuck together. Such operations generally involve reactions between chemical compounds and plasticisers and are complex.[104] In the assembly, account must be taken of the different coefficients of expansion of the materials.

Note: An unusual adhesive is that based on plasticised methyl cyano acrylate but unfortunately it polymerises rather quickly at 20°C and is expensive.

WELDING

Welding is a rather more effective technique than that in which adhesives are used. The joining together of two plastics items, such as films, by heating, compression and then allowing the joint to cool, is referred to as a homogeneous weld if only one material is involved and a heterogeneous weld if there are two or three different materials.

Ultrasonic Welding

The method most generally used is ultrasonic welding, which scours the surface before fusion; the melting point of the two components must be very close and the molten polymers must be compatible[105] (Fig. 34).

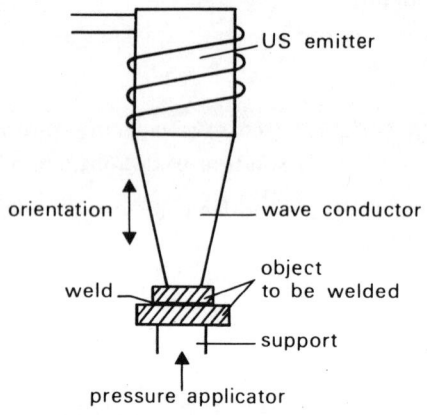

FIG. 34. Ultrasonic welding.

High-frequency (Dielectric) Welding

Welding by heating utilising the dielectric loss factor is only applicable at high frequency with highly polar plastics such as PVC and polyethylene terephthalate. The techniques cannot be used with polyethylene since it is not polar.

Welding by Hot Air Torch or by Infra-red

Materials can also be melted before welding by the use of a hot air torch or by means of an infra-red lamp (Fig. 35a).

Note 1: The surfaces must be cleaned with a solvent and alcohol after eventual straightening, particularly if different materials are involved, so as to ensure that the surfaces can be brought into close contact.

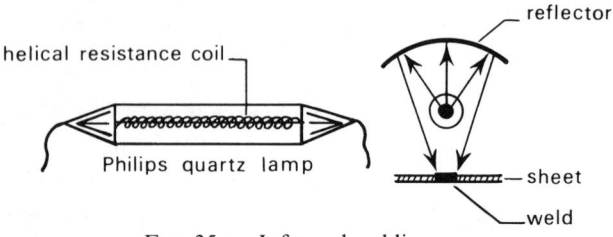

FIG. 35a. Infra-red welding.

FIG. 35b. Hot iron welding.

Note 2: Expansion must be taken into account.

Note 3: Thin films can be welded using a hot iron by passing it repeatedly over the material on cellulose paper (Fig. 35b).

Fixing and Joining
Films
They cannot be fixed to wooden frames directly using nails alone even when closely spaced. Polyethylene is most easily fixed using light wooden battens; the film is located on the framework and a lath, which should be of similar width to that of the supporting member, is then nailed down onto it. The advantage of having the lath of the same width is that it protects the film from degradation by sunlight at the point where it is stressed over the frame. Double headed nails are easier to remove when the film has to be removed.[106]

The film can also be kept in position over a framework by use of wide mesh netting of wire or plastics (usually nylon) stretched tautly between the support members. Tightened cables have also been used for holding film coverings on the supporting structure.

Where films need to be joined together and welding equipment is not available, the two edges with several inches overlap can be stapled together and then sealed with a special adhesive tape.[107] Widths of film can also be joined by stitching together with a 6 in overlap using nylon thread and a portable sewing machine.[108]

A novel method for fixing films involves the use of rigid profiles in expanded plastics such as PVC. The process (Fig. 36)[109] is intended for fixing films (low density polyethylene and others) on different types of frame and has two important features:

(i) the stresses of fixing are distributed along the whole length of the joint and not at isolated points;
(ii) installation and dismantling are quick and easy.

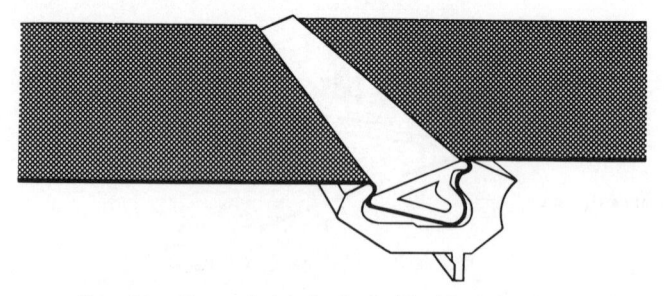

FIG. 36. Expanded polyvinyl chloride fixing joint.

Piping

Pipe systems can be made using several different types of fittings, although with polyethylene, compression fittings are most often used, and with PVC the joints are made by fixing the fittings by solvent welding. Where the pipe system is long, allowance must be made for expansion and contraction with varying temperatures, and expansion bends must be incorporated with mounting to allow movement of the pipe by means of suspension brackets (Fig. 37).

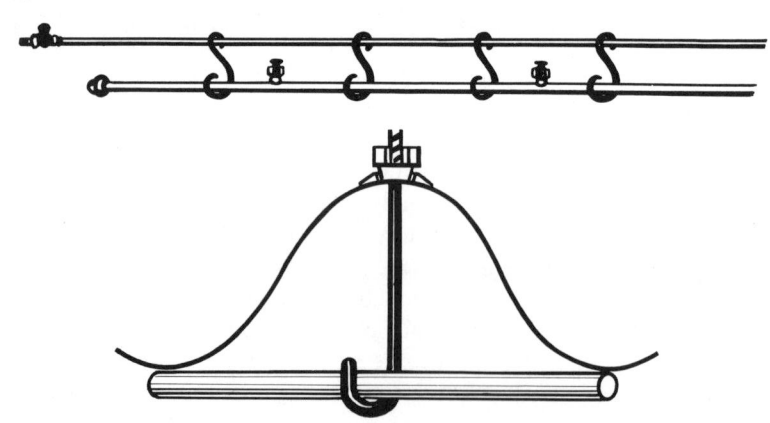

FIG. 37. Tube supports.

Sheets

Rigid plastics panels may be fixed in the same way as metal sheets, using ring washers and bolts, but in order to allow for thermal expansion, elongated holes should be provided for the fixing bolts. With corrugated sheets, the bolts pass through the ridges of the corrugations in order to avoid leakage of rainwater. Large panels of GRP, PVC, and PMMA can be fixed by the use of steel cables stretched over the panels and anchored into the ground. The flexibility of rigid PVC allows the construction of curved greenhouses.

The high coefficient of thermal expansion of PVC and PMMA calls for minimum fixing to a framework but this is not the case with GRP sheets which can be attached to wooden frames using galvanised screws.

LABELLING

This can be done with an adhesive paint but adhesion on polyethylene is rather poor unless the surface has been previously oxidised by using an electric discharge, by flaming, or with a chemical oxidising agent such as potassium bichromate. Ballpoint pens are not very effective and it is better to use felt marking pens or special paints on thermoplastics.

Chapter 5
Plasticulture in Practice

PATTERN OF CULTIVATION

The mechanisation of farming and crop growing has made extensive use of plastics materials which are employed in cultural techniques for the protection of plants against poor weather conditions and for the stimulation of their development.

Mulching has undergone rapid development as also has the use of low tunnels which in some areas, particularly in France, has increased considerably.

After windbreaks, protective nets and mulching, the development of protective structures has culminated in greenhouses where the artificial microclimate can be controlled precisely.

The cultivation of tropical crops calls for a special approach.[110]

Conventional ideas of cultivation are now being superseded by soil-less culture on an extensive scale, which now makes use of gullies formed from plastics with the nutrient solution being circulated through plastics pipes.[111]

MULCHING

In 1976, the mulched area in France was of the order of 35 000 ha, mainly using transparent polyethylene film, with a thickness of 30–50 μm, for plants such as melons, asparagus, strawberries and, to a lesser extent, tomatoes in the Vaucluse area, lettuce and endive in Roussillon, and cucumbers, beans, etc. In addition to polyethylene and EVA, other films in use are transparent or opaque PVC (plasticised) and thermal polyethylene. Black film is used extensively for strawberry cultivation, for humidity control and suppression of weed growth.[112,113]

Mulching of Vines

The plants mulched can be cuttings, either rooted or not. Any grafts must be plainly above the film.

The results of studies, obtained principally at the Institute for Wine in Nimes, France,[114-117] demonstrated the following:

increase in vigour and better root growth (Fig. 38);
earlier fruiting during the second year, but the difference from normal cultivation gets progressively less until the eighth year;
higher take with mulched cuttings (96%) as compared with 83% for the non-mulched;
reduction in the unfavourable effect of possible soil salinity.

Durability

It is desirable that film used in mulching, even if it is slightly degraded and embrittled, should be retained in position for several years, 4–5 years if possible. The use of such film reduces maintenance work such as weeding. This application requires the use of high quality polyethylene film produced

FIG. 38. Development of root growth under mulch: left, mulched plants; right, control. (By courtesy of R. Agulhon.)

FIG. 39. Mulching of vines. (By courtesy of R. Agulhon.)

from polymer with an MFI of 0·2, containing UV stabiliser for long durability (Fig. 39).

Strength
This is assured by a thickness of the order of 80 to 100 μm with a width of 1·5 m, the covered zone being about 1·1 m. The film is that classified as two-star quality in the CEMP standard.[101]

Resistance to Ageing and Weed Suppression
This is achieved by the use of opaque films, the most effective containing carbon black; transparent film is better for plant development during the first year but it does not prevent weed growth and its short life precludes its use in this case.

Mulching of Other Plants
Many experiences of mulching with low density polyethylene film have been described and some very encouraging results have been obtained.

Coffee and Cocoa
In the isles of the Gulf of Guinea weeding, which was previously essential every 2 months, has become unnecessary. In arid or semi-arid climates watering has been considerably reduced.[110]

Pears
In Portugal it has been observed that the increase during the first year in girth of the trunks of pear trees has been about 148% whereas the increase for the unmulched trees was only 100%.

Citrus Fruits
In Tunisia orchards of orange trees are mulched during the summer. A water saving of 30% has been achieved with an increase of yield of 50%.[118]

Bananas
The main interest in the mulching of bananas lies in the savings of water which can be achieved, particularly in those areas where there is a constant shortage, such as the Cape Verde Islands.[119-121]

Nursery Stock
Observations of improved growth have been recorded on young trees intended for fruit and forestry plantations.

Maize
According to the work carried out by Ballif and Dutil[122] the effects of a transparent plastics mulch on maize grown in the chalky soil in Champagne, France, are as follows:

seed germination takes place about 15 days after planting instead of 20 days;
more regular and vigorous growth of the plants occurs (Fig. 40a);
the crop can be harvested about 3 weeks earlier;
the yield increases by about 25%.

Problems encountered with the collection and disposal of the film have now been largely overcome by the use of photo-degradable film (Fig. 40b).[123] Trials with films which start to degrade after 30, 60 and 90 days have shown that, for growing maize, the best results are obtained with the material which begins to degrade after 90 days. In order to facilitate the passage of the plants through the film, it can be perforated at the time of sowing, but in the US, slit film (in bands of about 2 in) is used extensively.

86 PLASTICS IN AGRICULTURE

FIG. 40a. Effect of mulch on growth of maize. Sown 21st April, photographed 4th June; control (unmulched) on the right. (By courtesy of CdF Chimie.)

Melons

Type of film. Except in those areas where weed control is a problem, growers generally prefer to use grey or transparent films rather than black film because the crop is earlier and the yield is greater.

Perforation of the film. This is essential, to avoid the pool of water which would otherwise form under the heavy weight of the fruit.

Cultivation of melons in Provence. This involves a lot of work in February, with the application of a mixed organo-mineral fertiliser and the sowing, between the 10th and 15th April, of six to twelve seeds together in a group spaced 0·8 m apart in rows separated by 2·5 m. The film is then set out and between the 2nd and 5th May it is slit to the right and to the left of the

Fig. 40b. Maize mulched with photo-degradable low density polyethylene film. Control (unmulched) on the right. (By courtesy of CdF Chimie.)

group of seeds, then, a few days later, a further slit is made towards the plant and the seedlings are thinned out to leave two to three in the group. It would be preferable to make the openings with a heated perforator rather than a knife because of initiating tears. It is essential that the leaves are not in excessive contact with the film to avoid scorching the plants.

Strawberries

Type of film. Black polyethylene with a thickness of 50 μm and a width of 1·1 m is used, so as to have a mulched strip of 0·8 m.

Note: In California the film mainly used for this application is transparent polyethylene.

Technique. This can be done by establishing a centre row of parent plants which are increased on the two sides by rooting the runners or by planting two or three rows of established plants at distances of between 0·2 to 0·4 m, with regular removal of all the runners.

The earliness of planting and the time of harvesting go together so that the development of multi-cultivation during the year with two crops annually gives better utilisation of the ground and reduces the risks of contamination by viruses. Consequently it is becoming general to use plants of the variety 'Frigo' which are planted under plastics film and then discarded after the harvest. The plants with incipient flower buds are selected at the end of winter and then kept in a cold chamber at about −1 °C, thermetically sealed in polyethylene bags. Planting takes place very early, between the beginning of June and 15th August, which gives the maximum yield at the end of the first year; if the plants are grown on for a second year, it would fall by 20 to 50%.

In California, drip irrigation is frequently used in conjunction with mulching for the cultivation of strawberries. A further development is the production of equipment for the fumigation of the soil, the laying of the film and the installation of the trickle irrigation system all in one operation.

Extensive work has been carried out on the development of cultural techniques for strawberries both in France and in the USA.[124-128]

General Conclusions
Mulching with the help of plastics films has, for over a decade, played a major role in plant cultivation by creating at the soil surface some measure of *mechanical protection* and a *microclimate* which is favourable from the aspects of temperature distribution and retention of humidity, surface fermentation, and the supply of carbon dioxide to the stomata of low plants.

It can be used very effectively with trickle or drip irrigation, particularly for the cultivation of vines and strawberries, thereby encouraging the lateral development of the root system.

TUNNEL STRUCTURES FOR SEMI-FORCING

Objective
Tunnel structures can be used for semi-forcing so as to grow, without heat, early or late crops with an increase in the yield.

Films Used
Low density polyethylene is mainly used, with some use of PVC and EVA. The properties of the films employed are summarised in Table 13.

Comparison between Polyethylene, PVC and EVA
Low density polyethylene and transparent plasticised PVC (28% dioctyl phthalate) have comparable qualities as regards flexibility, lightness and radiation permeability to short-wave infra-red light which penetrates into the interior of structures and heats up the soil and the plants.

During the night, polyethylene is equally permeable to the long-wave

TABLE 13
PROPERTIES OF FILMS USED IN TUNNEL STRUCTURES

Property	Low density polyethylene	Crystal PVC (28% dioctyl phthalate)
Density (g/cm^3)	0·92	1·3
Tear strength at 20°C	50–80	50–60
Temperature range	−40 to 70°C	−10 to 50°C
Durability	1 season	1–3 seasons
Transparency (0·38–0·76 μm)	70–85%[a]	80–87%[b]
% Transmission between 0·24–2·1 μm	80[a]	82[b]
% Transmission between 7–35 μm	80[a]	30[b]

[a] For a sheet of 80 μm thickness.
[b] For a sheet of 100 μm thickness.

infra-red radiation emitted by the soil, thereby giving rise to a high thermal loss. Polyvinyl chloride differs in that it is impermeable to long-wave infra-red, so that heat losses during the night are less and the temperature is therefore higher under PVC than under polyethylene. When using polyethylene a temperature inversion becomes possible, i.e. in the region of 0°C the temperature is lower inside a structure than outside. Crops grown under PVC are therefore earlier than those grown under polyethylene.

This problem has been overcome in France by the introduction of 'infra-red' polyethylene which improves the permeability to resemble that of PVC so that comparable results are obtained with the two materials.

Ethylene–vinyl acetate film has only recently become available. It is formed by the co-polymerisation of ethylene with vinyl acetate and it has improved permeability characteristics in the infra-red spectrum so that it competes with PVC in the production of early crops grown under cover. The thermal behaviour of EVA depends on the vinyl acetate content and it would appear that the optimum level is between 12 and 15%.

Types of Low Tunnels
Several different types of low tunnels exist; they are generally about 1·25 m wide and 40–50 cm high. They provide cheap protection and are particularly useful for covering low crops such as melons, strawberries, lettuce and carrots (Fig. 41a).

The most widely used is the '*Nantais*' *tunnel* with double hoops with the film held between them so that it can be slid upwards to allow ventilation.

Fig. 41a. Low tunnels for growing peppers in Israel. (By courtesy of H. R. Spice.)

Fig. 41b. Tunnels using two films.

The only disadvantage of this type of tunnel is the labour required for adjustment of the ventilation.

Tunnels with single hoops are set up either by hand or mechanically—the film is stretched out over the hoops and then the two edges are buried. Ventilation is introduced simply by making holes in the sides, depending on the climatic conditions.

In the USA, two films are used over metal hoops; these are fixed at the top by clips to a steel wire which is stretched along the length of the hoops (Fig. 41b). The plants can grow upwards between the two films by the opening of the clips; this type of low tunnel can be easily ventilated and maximum ventilation obtained by removing one of the films.

Coverings of Perforated and Permeable Films

Film coverings without the need for support can be used for semi-forcing by employing perforated and permeable films; these give protection to early crops grown in the open in the spring. This type of covering has similar advantages to the low tunnels, i.e. earlier crops, by 8 to 10 days; better quality produce; staggering of production; and protection against birds.

Two types of film are used, either polyethylene of 50 μm thickness and having 500 to 700 perforations of 7–8 mm diameter per m^2 (Fig. 42) or films which are cut to form slits; these slits are closed when the film is first laid out over the plants but they open progressively as growth takes place.

This technique gives excellent results with certain crops such as potatoes.[129] It appears, however, that the direct contact between the film can cause necrosis of the leaves.

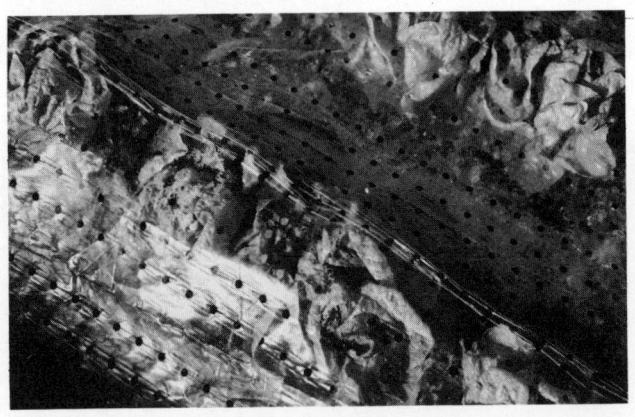

FIG. 42. Perforated film for crop forcing (Mandylene 500 holes). (By courtesy of J.-C. Garnaud of the CIPA.)

FIG. 43. Mulched strawberries grown under low tunnels.

Use of Mulch in Protective Structures
This is associated with mulch films in low tunnels and is often used with crops such as melons and strawberries (Fig. 43).

PLASTICS GREENHOUSES

General
The greenhouse is a structure with a covering and walls, either flat or curved, transparent or translucent, in which it is possible to maintain an atmosphere more or less conditioned as regards temperature, humidity and radiation energy, so as to encourage crop earliness—and sometimes to retard it—improve the yield, safeguard the crop and make more effective use of water.

It is essential that the control and possibly the variation of the artificial climate thus created are suitable and seasonable as a result of using satisfactory *automation* and that the manual or mechanical operations are made easier by the topography and the arrangement of the sites.

Note: The types of structures and the designs of greenhouses are reviewed in references 130–136.

Properties of Films
Density of Material (g/cm^3)
This is low (about 0·92 for polyethylene and less than 1·5 for PVC) while it is

about 2·5 for glass, which makes up only about one third of glass-reinforced polyesters. This means that the framework necessary can be lighter for plastics greenhouses than for glass-houses and also that the shading zones will be less in the former than in the latter. The light weight of construction of plastics greenhouses and their resistance to impact make them easy to move for crop rotation while the rounded form helps to make them air-tight.

Transparency to Solar Radiation
This is of the same order as for glass but there are certain differences with the various materials. The permeability leads to effective heating during the day and this is followed by rapid cooling at night, particularly with polyethylene, although this effect is, in fact, compensated in practice by the presence of condensed water on the internal wall or by the use of a double wall of film. General data on permeability to radiation and thermal conductivity of films are given in Table 14.

Heating
This is more expensive for a glass-house than for a double-walled plastics greenhouse, since the heat required is 10 and 7 cal per m² respectively for each 1 °C difference in temperature between the exterior and the interior.

Air Humidity
Because of their low permeability to water vapour (discussed in Chapter 2) plastics greenhouses permanently maintain a higher degree of humidity resulting from evapotranspiration.[137]

Ventilation
This is more important in plastics greenhouses than in glass-houses. It is necessary to ventilate early in the morning before the temperature rises, whether the structure is heated or not.

Static ventilation systems are simple and relatively cheap but they are often inadequate and require excessive labour. Forced air systems are very effective, and should be based on a ventilation rate of 30 to 40 air changes per hour. Low-speed fans, giving a high volume flow at low pressures, are recommended.

Air-tightness
This is the main advantage of plastics greenhouses, particularly those with flexible film coverings, with superior air-tightness as compared with glass-houses. With all other things being equal, in a comparison of the two types,

TABLE 14
PROPERTIES OF MATERIALS USED IN GREENHOUSE COVERING

Property	Low density polyethylene $(CH_2-CH_2)_n$	Crystal PVC (plasticised) (PVC + stabiliser = 72% Diethyl phthalate = 28%)	GRP (Polystyromaleate phthalate of ethylene glycol)	Glass
Transparency 0·38–0·76				
Average % transmission between 0·24–2·1 μm	70–85 %[a]	80–87 %[b]	85 %[c]	87–90 %[d]
Average % transmission between 7–35 μm	80[a]	82[b]	60–70[c]	85[d]
Thermal conductivity (cal/cm²/sec/°C)	80[a]	30[b]	0[c]	0[d]
	7×10^{-4}	4×10^{-4}	$1 \times 10^{-5\,c}$	$24-27 \times 10^{-4}$

[a] For film of 0·08 mm thickness.
[b] For film of 0·1 mm thickness.
[c] For sheet of 2 mm thickness.
[d] For a thickness of 2·7 mm.

the automatic ventilators of a polyester greenhouse will close one or two hours later than the ventilators of a glass-house.

Work carried out in France[138] has shown that the soil spontaneously gives out carbon dioxide in the absence of plant life. In a greenhouse which is perfectly air-tight the concentration of the gas reaches a level of 7 times that of the exterior. It is recommended however that the air should be cooled (from 35 to 26 °C in the experiment reported), and this assures the earliness of the crop and twice the yield in the case of spinach.

Resistance to Hail
Contrary to the case of glass-houses, the resistance of plastics greenhouses is such that it is not necessary to take out insurance against this hazard, except for those greenhouses constructed of rigid PVC.

Types of Plastics Greenhouses

Ultra-violet stabilised polyethylene which has a life of at least two seasons is used almost exclusively as the covering material for greenhouse structures in *flexible plastics*; the standards which exist in France and the UK ensure that the films have good mechanical properties. Polyvinyl chloride films which give good agronomic results are available only in relatively narrow widths; they tend to pick up dirt and the mechanical properties deteriorate after 1 year's exposure but this factor can be overcome by reinforcing the film with a mesh of nylon or polyester. In Japan, special PVC films which have been treated to prevent dirt pick-up are now used, and similar developments are now in progress in Europe. In certain countries such as Japan and Finland, EVA films are already widely used, where double-walled structures are based on EVA film with a 4-year life (220 μm thickness for the exterior skin and 100 μm for the interior).

Woven polypropylene has also been evaluated but rejected because of the high rate of dirt pick-up which considerably reduces the light transmission.

Note: These structures are generally fitted with less sophisticated conditioning equipment than greenhouses constructed using rigid sheets.

Greenhouses in *rigid plastics* are generally constructed using sheets of glass-reinforced polyester or rigid PVC and can have an arched form. The equipment which is fitted for heating, ventilation and irrigation is usually of up-to-date design.

FILM GREENHOUSES

Structures with a Wood or Metal Framework
Wooden Frameworks
From about 1960, film of 200 μm thickness has been attached to a wooden structure which has become lighter and lighter. It has proved necessary to hold the film in position with wide mesh metal netting or wire strands, in order to prevent it from flapping in the wind. This simple form of construction can be used in climates such as in the Mediterranean regions, where there is usually no heating, for growing early salad and tomato crops, or where wood is cheap (Fig. 44a).

FIG. 44a. Protective structure constructed by a grower in Spain.

Metal Frameworks
These are generally constructed from galvanised metal tubes, the dimensions of the tunnels being of the order of 5–7 m in width and 2–2·5 m in height. The assembly of the metal framework can be quickly carried out. Polyethylene film is now available of such width that tunnel structures can be covered with a single sheet. A type used extensively in France is covered with film in the transverse direction, each width overlapping the next by about 0·30 m so that the tunnels can be ventilated by creating openings between the sheets (Fig. 44b).

Structures with a Double Skin
This type is covered with a double thickness of polyethylene using a metal framework; in France they are 8·5–9 m wide and 3·2–3·5 m high.

FIG. 44b. Tunnel greenhouses built by 'Filclair'.

Ventilation is obtained with framed ventilators or with fans which can be used in conjunction with a heating system. Numerous other types of structures have been described.[139]

Static and Forced Air-separated Double-layer Plastics Greenhouses

The increased cost of fuel for heating has encouraged studies to reduce the heat loss from plastics protective structures and this can be most readily accomplished by use of double-layer film coverings separated by an air space. It has been concluded[140] that the distance between the layers should ideally be 3·8–5 cm, to maintain a dead air space for maximum insulating effect. Many of the construction details and inflation requirements were worked out by Roberts at Rutgers University.[141,142] Comparative studies have been reported[143] on static and forced air-separated double-layer plastics greenhouses. In the former, the roof and sides were covered with two layers of polyethylene film (100 μm) which were kept apart by 5-cm wood strips. The latter had two layers of film kept apart by air injected from a blower delivering air at 1·3 cm static water gauge pressure; the maximum distance between the two layers at the centre of the roof was about 75 cm, and between the side walls about 12 cm. Measurements of heat losses showed a saving of about 40%, but the framed structure using 5 cm wood spacers was more expensive to construct than the double-layer forced-air house (Fig. 45).

Other authors[159] have studied yields, cost of construction and light

FIG. 45. Greenhouses covered with two layers of film with air inflation to conserve energy. (By courtesy of J.-C. Garnaud of the CIPA.)

transmission for inflated double-skin and single-skin structures. It has been shown that the double film reduces light transmission by 10–14 % but since the structural strength is greater, fewer supporting members are required, which compensates for the use of two skins. In Israel, where very good light conditions prevail, very little difference in luminosity inside the two types of structures has been recorded.

Air-supported Greenhouses

These are usually semi-cylindrical structures, maintained in shape by using air pressure slightly above normal. A simple door is quite adequate but air locks are often provided and the air pressure is supplied by high-volume, low-pressure fans.

There are few problems involved if semi-cylindrical shapes are chosen and if the width at ground level is restricted to about 9 m—the width which can be conveniently covered with sheeting 11 m wide: polyethylene film of 125 μm thickness is normally used. The fans should be capable of maintaining a pressure of about 12·5 mm water, in order to provide sufficient stability in high winds, but under normal weather conditions pressures as low as 2·5 mm are satisfactory. Higher pressures may cause excessive stresses in the envelope. The structure is therefore fitted with a ventilator consisting of a hinged flap which is counterweighted so that it will remain closed at the required pressure but open when a high pressure develops. The ventilation required may be from zero up to 4 m^3 per min per m^2 of covered area, depending on the crop, the stage of growth and the time

FIG. 46. Air-inflated polyethylene structure. (By courtesy of H. R. Spice.)

of year. There are numerous other studies on this form of construction[144-150] (see Fig. 46).

Advantages
Absence of structural supports gives improved luminosity and access for mechanical equipment.
Very large areas can be covered fairly cheaply (PVC air-supported greenhouses are in commercial production in the USA, measuring 428 ft long, 100 ft wide and 20 ft high).

Disadvantages
The structure collapses in the event of an electricity failure but this can be minimised by using an air-inflated roof on a simple framework on which high tensile wires are stretched above head height.
Supports have to be provided for growing crops such as tomatoes and cucumbers.
Although extensive work has been reported, acceptance of this type of structure for greenhouses has not been widespread, due mainly to the need for stand-by electrical equipment and, in some cases, inadequate stability under heavy wind and snow loadings.

Development of Plastics Greenhouses in the UK
Development of plastics covered structures began in 1968, chiefly at the Lee Valley Experimental Horticulture Station. A prototype single-span plastics covered tunnel 4 m wide and about 40 m long was covered with a single sheet

FIG. 47a. Single-span plastics-covered tunnels fitted with fan ventilation at Lee Valley Experimental Horticulture Station. (By courtesy of the Ministry of Agriculture, Fisheries and Food.)

of UV-stabilised polyethylene film (7·5 m × 45 m). The framework was made up of galvanised metal tubing consisting of hoops, ridge pieces, diagonal braces and foundation tubes for receiving the ends of the hoops. Tensioning wires were also fitted, and the entrances at the two ends were of timber construction. The film was stretched over and secured to the framework, the edges being buried in a shallow trench running alongside the structure[131] (Fig. 47a). Continuing development ultimately led to work on multi-span tunnel greenhouses, which commenced in 1971. The original

FIG. 47b. Double-span structure under construction at Lee Valley Experimental Horticulture Station. (By courtesy of the Ministry of Agriculture, Fisheries and Food.)

prototypes consisted of double- and triple-span structures clad with polyethylene film. The double-span type (16 m × 36 m) was provided with three lines of stanchions (2 in galvanised steel tubing) and the framework was formed of hoops using 1 in tubing. The height to the gutters was 2·5 m and the pitch of the curved roof was 45° at the gutter (Fig. 47b). Subsequent developments have led to improved forms, and recommendations are contained in a special publication.[132]

An eight-span structure has since been erected with an overall area of 63 m × 36 m (Fig. 47c).

FIG. 47c. Multi-span structure at Lee Valley Experimental Horticulture Station.

Examples of Cultivation under Flexible Plastics Structures
This is often referred to as 'semi-forcing' since the crop is grown under low or walk-in tunnels without resort to heating.

Melons
Experiments have been carried out at the Centre Agronomique at Montfavet in France on a comparison between cultivation in a glass frame and under low tunnels clad with polyethylene. Crops have also been grown under glass cloches and in frames clad with PVC. It has been observed that the structure must be firmly closed at night to keep the heat in and that, at the flowering period, good ventilation must be maintained.

If growth is too luxuriant, the protection must be removed, taking care that the plants are hardened off gradually.

Tomatoes

Trials have been carried out at the Lee Valley Experimental Horticulture Station on the growing of a number of different varieties of tomatoes in unheated tunnels.[151] In the initial stages some problems arose because of inadequate ventilation but improved results were obtained in subsequent years by paying more attention to this aspect. Studies have also been reported on the cultivation of this crop in Poland.[152]

Lettuce

This crop has also been grown in unheated tunnels at Lee Valley[151] using a number of different varieties over practically the whole year. Studies carried out in Belgium[153] indicate that better lettuces are obtained if the plants are covered for only a limited period and 21 days must be considered as the maximum.

Carrots

Cultivation in plastics tunnels has been carried out by a number of workers.[151,154,155]

In France, in the Nantes region, early carrots are sown in the autumn after soil preparation which consists of breaking down the soil and then adding the fertiliser, when the opportunity arises, on the area of the beds for each plot. These beds, which are at right angles to the axis of the main plot, are 1·25 m wide and are separated by paths 0·5 m wide. The inorganic fertiliser incorporated before sowing is in addition to the spreading of manure and account is taken of the fertiliser added as a top dressing in the previous spring.

Organic fertiliser (manure)		40 ton/acre
Inorganic fertiliser before sowing	N	12 to 16 kg/acre
	P	20 to 32 kg/acre
	K	40 to 60 kg/acre
Top dressing of inorganic fertiliser (in one or two applications)	N	6 to 10 kg/acre
	P	12 to 20 kg/acre
	K	24 to 40 kg/acre
Total inorganic fertiliser	N	24 kg/acre
	P	40 kg/acre
	K	100 kg/acre

Sowing is carried out between 25 October and 15 November with 16 to 17 rows for each bed spaced 7·5 cm apart using about 1 g of seeds to the square metre and less if they are pelleted, which saves time on thinning. A roller is used to consolidate the beds, the seeds being covered with about 1 cm of sand. Each bed is covered with a line of cloches or a tunnel.

Weeding is either done by hand or with herbicides, except where the carrots are grown with other plants such as lettuce, radish and lamb's lettuce.

In January or February the crop is watered and if necessary a top dressing of fertiliser is applied. The crop is uncovered on 15 March and another top dressing is applied with a spray against aphids and carrot fly (which is on the wing in the latter part of April). The crop is dug on about 15 May and the leaves are removed. After washing the carrots are sorted and made up into standard packs of 13 kg.

Sweet Peppers

Trials were carried out at Lee Valley[151] on the cultivation of 12 different varieties to determine those most suitable for growing in unheated tunnels. The seed was sown in the middle of March and after pricking out into pots the seedlings were grown on in a heated glass-house and then planted in a tunnel during May. The plants were fed with high nitrogen tomato fertiliser which was given at most waterings. The crops were harvested from 17 June to 17 November; five varieties appeared suitable for tunnel cultivation with an average yield of about 20 tons of fruit per acre.

Strawberries

Mulching is sometimes combined with semi-forcing for melons, and this is normal for strawberries which show up with earlier crops and better yields while also reducing the maintenance work.

Crop earliness seems to be approximately proportional to the volume of air within the structure—1 m^2 of plant area corresponds to a height of 0·4–0·5 m. In addition, with a film covering of the tunnel, bearing a layer of condensed water, frost damage is often prevented if the volume of the tunnel is large enough. Plenty of ventilation between flowering and setting is essential, in order to reduce the amount of foliage and to help fertilisation, particularly if mulching has been used which prevents the spread of *Botrytis* that can occur in a tunnel in a favourable microclimate.

Endive

This is a salad crop which will tolerate temperatures down to −10 °C. The seedlings are pricked out at 6–8 cm both ways and as soon as they have four or five leaves they are planted 0·4 m apart in rows spaced at 0·4 m with a distribution of 1 kg of ammonium nitrate per are (100 square metres). The plants are covered with a tunnel 0·5 m high as soon as white frosts begin; this can then be used in the spring for strawberries and radishes.

Tobacco

Mulching with plastics film is used for nearly 90 % of the seed beds which are covered with low tunnels of 1·2 m diameter. Pre-germination sowing of the seed is done in sachets made of low density polyethylene maintained at 20–30 °C in a humid atmosphere. Sowing is then carried out on a warm bed, the seeds being mixed with 100 parts of ash. The requirement is for 1 m^2 of seed bed for every are under cultivation; since this is about 20 000 ha this corresponds to 200 ha of seed beds.

The plant density is 1000 plants to the square metre. They are top dressed with compost through a sieve. Some sand is added and they are then re-covered with polyethylene film.

The variety Paraguay has been replaced by a hybrid, P139N, because of its greater resistance to mildew. Deflowering is carried out in order to encourage leaf growth and is done by pouring a vegetable oil over the first

Fig. 48. A battery of tobacco driers.

flowers which appear at the top of the stem. The oil then runs down on the buds, which form symmetrically on both sides of the stem at the base of the leaves, and which are thereby destroyed. Drying can be done in a drier covered with asphalted sheets and protected on the sides with polyethylene film or other plastics. Greenhouse driers (as in western France) or special tunnels (as in south-eastern Alsace) are used. The hoops are PVC tubes (36/40 mm × 6 m) at 1·3 m spacings and are connected together with ropes (Fig. 48). The hoops are set in steel tubes which are partially buried, the film used being low density polyethylene of 100 or 150 μm thickness and of 7 m width.[156]

Other Crops
The advantages of cultivation of vegetables and salad crops in plastics tunnels have been extended to cover such crops as cucumbers,[157] cauliflowers,[158] and aubergines and celery.[151] There is now extensive production of flowers such as carnations, roses and chrysanthemums in large tunnels.

Conclusions on Semi-forcing in Plastics Tunnels
The main advantage is that crops can be produced about 2–3 weeks earlier than in the open, with improved yields, by protecting them against frost and wind. Installation is easy, particularly if mechanical equipment is used and the investment is relatively low.

Use of Low Tunnels inside Larger Structures
There is an increasing effort to gain additional time on the earliness of the crop by advancing the date of planting in unheated tunnels and protective structures. In order to guard against the risks of frost, the plants are temporarily protected by low tunnels after setting out. In areas where light is not a limiting factor, temperatures are obtained which are 3–6 °C higher than the outside.

RIGID PLASTICS GREENHOUSES

The materials which are generally used are sheets of fibre-reinforced polyester, rigid unplasticised PVC or PMMA.

Reinforced Polyesters
Composition
These are usually based on polyester resins reinforced with glass fibres; flat or corrugated sheets are available, the properties depending on the

composition of the resin and the amount and distribution of the fibres, which can be aligned or random. The composition (e.g. the proportions of maleic anhydride and phthalic anhydride) has a marked effect on the mechanical and chemical properties of the resin; the presence of tetrachlorophthalic acid in place of phthalic acid increases the refractive index and the use of methyl methacrylate in place of styrene in a resin can lower the diffusion power of the product from 50% to 10%.

Optical Properties
The polyesters are, in general, only slightly transparent to UV radiation (at wavelengths of 0·3–0·38 μm) and the penetration is further reduced and even eliminated by UV absorbers. Visible radiation is markedly diffused; the addition of 15% PMMA increases the transparency and stability of the finished product considerably. In all cases, transparency to solar infra-red radiation (at wavelengths of 0·76–2 μm) is lower than that of glass. All things considered, since the total transmission is lower than that of glass, glass-reinforced polyesters give rise to a reduced temperature build-up; this is of greater advantage in the south of France than in more northerly climates. However, because of the opacity of this class of materials to infra-red radiation emitted by the soil (at wavelengths of 2·5–35 μm) the 'greenhouse effect' is produced.

Rigid PVC
General Performance
Rigid PVC is available as flat or corrugated sheet. The composition of clear rigid materials can vary appreciably according to the stabilisers and lubricants which are used. In general this material and GRP may be used for cladding with roughly the same advantages but they certainly cannot be interchanged.

The opacity of transparent sheet increases with exposure to outdoor weathering and the development of a yellow to dark brown colour reduces light transmission to such an extent that replacement ultimately becomes necessary. Degradation is accelerated in those areas where the sheet is in close contact with the supporting structure and consequently local hot spots are created. Clear rigid PVC does not suffer the surface roughening which occurs with GRP.

Optical Properties
The light transmission curves of glass, GRP and rigid PVC are effectively the same as regards the light reaching the soil surface. The light diffusivity of

rigid PVC is inferior to that of GRP. The light transmission of rigid PVC is given[160] as 70–80 % compared with 80–90 % for GRP. The material is also opaque to long-wave infra-red radiation.

The clarity of different products can vary appreciably and there has been considerable development in grades suitable for agricultural applications and which are guaranteed to be satisfactory for at least 5 years before replacement is needed.

Polymethyl Methacrylate (PMMA)

General Performance

Polymethyl methacrylate has been used for a long time in applications requiring a rigid transparent plastics material. It has excellent outdoor weathering characteristics and can be readily formed into curved shapes. It does undergo some yellowing on prolonged outdoor exposure but this can be reduced by the incorporation of UV absorbers.

Light Transmission

This is better than for rigid PVC, the light transmission being in the order of 92 % for clear panels. It has been shown that PMMA has a higher transmission than glass in the infra-red region of the spectrum. The energy transmission in that sector of the spectrum which is accepted as giving maximum growth response (between 0·4 and 0·75 μm) is about 10 % greater than for glass and is also higher than for GRP and rigid PVC.

Greenhouses in the Traditional Form

The materials already described have been used for greenhouse structures of conventional glass-house design, with the metal framework structure being more frequently adopted than the wooden framed structure with spans of 6, 9, 12 and up to 16 m.

Because of the high coefficients of thermal expansion, PMMA and rigid PVC must be fixed to the framework at the minimum number of points and special techniques, such as the use of cables, have been employed. The lower thermal expansion of GRP does not call for the same critical methods of fixing. There has been continuing development in grades of the rigid plastics materials for use in greenhouses and from time to time comparative studies are reported.[161]

Work has been carried out at Efford Research Station, Lymington, in the UK, on the use of various types of rigid plastics. These include two grades of PVC, GRP and PMMA. Although PMMA had superior light transmission and the GRP used showed the most rapid loss, the differences in light transmission values did not exert such a great effect upon the crop growth as might have been expected.[162]

In Sweden, work has been carried out since 1964 and it is of interest to note that the best rigid glass-reinforced polyesters had a reduction of only 1–3% in light transmission after 5 years; considerable variation was experienced with different grades of rigid PVC.[30]

Rigid PVC has been used in the UK on structures with conventional pitched roofs but with no side walls.[163]

In the USA, glass-reinforced polyester panels are finding increasing usage. Although the cost of the sheet is greater than that of glass, the lighter weight can lead to substantial savings in the cost of the structure so that the final installation cost is competitive with the cost of installing glass.[164]

Greenhouses with Arched or Curved Roofs
The use of rigid sheets which can be shaped or flexed into a curved form presents some advantages over the conventional design:

1. *Maximum diffusion of light*, due to having a good radius of the arch which can be obtained using flat sheets but not with corrugated sheets.
2. *Simplification of the framework*, by the use of galvanised steel tubes or laminated wood with sufficient span between the posts.
3. *Maximum tightness* by the use of a single sheet over the whole width.
4. *Fuel savings*. The air volume is less than that in a multi-span greenhouse as also is the heat exchange surface with the exterior, and the thermal bridges are fewer. Many designs have been described and these have been reported by Keveren.[165]

Other Types of Greenhouses
Mobile Greenhouses Constructed from GRP or Acrylic (PMMA) Sheet
A greenhouse covering 1000 m² and made up of spans 9 m wide is mounted on rails; it weighs only 14 tonnes and can be readily moved as a single unit. Figure 49 shows a mobile greenhouse made up from extruded polymethyl methacrylate (Perspex).[166]

Self-supporting Greenhouses
Structures of this type were developed in Germany in 1961 using glass-reinforced polyesters. Wide span shell-like structures could be produced

Fig. 49. Trailer Greenhouse clad with 'Transpex' extruded acrylic sheet. (By courtesy of Imperial Chemical Industries Ltd, Plastics Division, and made by E. J. Frost of Eccleshall, Staffs., UK.)

from individual elements with U- and L-shaped edges which gave the individual segments great stability and enabled greenhouses to be constructed with no supporting structure. These particular structures were stable against the specified wind and snow loadings and had an economic advantage since, in a conventional greenhouse, the support structure could account for 65% to 75% of the total constructional costs. The components for these houses were supplied as prefabricated units which were light in weight and easily assembled.[167,168]

Tower Greenhouses

These have been produced in Austria especially to overcome the problems of land costs in urban areas and to make maximum use of the sunlight. The tower is 42 m high and 8 m in diameter. The plants are conveyed along on vertical conveyors so as to be subjected to optimum solar radiation and are watered periodically by passage through a bath containing a nutrient solution at the bottom of the cycle. Originally the tower was made from glass, but at a later date the panels were replaced by corrugated GRP. Environmental conditions were investigated so as to produce maximum yields throughout the year in an industrialised horticultural process.[169]

DIFFERENT SYSTEMS OF HEATING

Central Heating with Steam or Hot Water

This is the classical method of heating, in which pipes are run along the walls of the greenhouse under the eaves and along the supporting framework at a height of 50 cm for salad crops and 10 cm for tomatoes. Movable pipes can be used in the centre of the spans.

Warm Air Heating

Warm air generators are commonly used; these are placed inside the greenhouse and the combustion products are led away to the outside (Fig. 50a). The distribution of the warm air can then be arranged in several different ways.

1. Air ducts in the upper space of the greenhouse: The heated air is delivered from a number of outlets.
2. Ducts in the walk-ways.
3. Perforated thin-walled polyethylene tubing: This is about 30 cm in diameter (thickness 120–150 μm) and has holes of 10–15 mm at 50 or 60 cm along the entire length. The polyethylene can be located either in the upper parts or 50 cm from ground or soil level between the plants (Fig. 50b). The advantages of heating with forced warm air using perforated thin-walled polyethylene tubing are: a uniform

FIG. 50a.   Gas heater used for greenhouse heating.

FIG. 50b. Greenhouse provided with heating using polyethylene film tube.

temperature distribution is obtained by convection from the surface and streams of warm air at low velocity, thereby reducing the convection from the walls of the greenhouse. This ensures that there is no excessive local build-up of humidity which can cause disease. The mobility of the system and the reasonable investment cost have encouraged its use.

Air–Soil Heating

This is used in horticulture, particularly where the soil is frequently replaced. The tubes (2 cm in diameter) through which warm water circulates are buried at a depth of 25–30 cm and spaced at 20–25 cm apart. They are covered with a 5 cm layer of sand and then with earth; the temperature inside the tubes will be of the order of 35 °C and the soil at the level of the roots 25–27 °C. If more intense heating is used (e.g. hot water at 100 °C) tubes of polypropylene or an ethylene–propylene copolymer are used (Fig. 51).

Flexible Tube Soil Heating

This technique is a recent development which may utilise waste heat from electric power installations. Initially, tubes of layflat polyethylene of about 200 μm thickness were located between the rows of plants and warm water at 25–30 °C was circulated through them; the tubes covered more than 50 % of the total growing area. After the initial trials, the space between the plants proved insufficient and the tubes were then formed into mattresses

FIG. 51. Polyethylene tubes for warm water circulation.

FIG. 52. Close-up of soil heating using polyethylene film 'mattresses'. (By courtesy of J.-C. Garnaud of the CIPA.)

with holes in them so as to allow the plants to grow through (Fig. 52); this allows normal planting densities to be adopted and ensures more uniform heat distribution around the roots.

The warm water may be obtained from a geothermic (underground) source and also by heat exchange with the cooling water from nuclear reactors; the system may also utilise solar energy. It is not unusual to find that during very sunny days water which enters the tubes at 27°C rises in temperature due to heat absorption by the black polyethylene.[170,171] This heated water can be stored and then used at night for additional heating.

Note: An article on the various techniques of heating plastics greenhouses has appeared.[172]

VENTILATION

Ventilation may be necessary during winter as well as summer. To provide satisfactory cooling in summer a ventilation rate of 5–7 ft^3 of air per min per ft^2 of covered area has been recommended for plastics film *greenhouses*. Generally ventilation is obtained by means of openings at the sides or at the ridges (Figs. 53a and b) but houses are provided with forced draught reversible ventilators in the gables, the air being renewed every 4 or 5 min. Where sections are removed to permit ventilation the opened area should represent from one-tenth to one-eighth of the total cover surface.[173] One of the gables is made up of wet film fibre pads and the other carries the ventilators, by which means a lowering in temperature of 4–5°C compared with the exterior can be obtained. The temperature can also be reduced by a water trickle on the roof.

Note 1: In an automated American system, a ventilation fan creates a vacuum inside the greenhouse when the control switches on the cooling; the cold air is sucked in through the flexible tubing connecting one of the gables to the other.

Note 2: The control of heating and ventilation is becoming more developed with the progress in automation. A system developed at the Kentucky Agricultural Experimental Station[174] consists of a continually running fan providing ventilation, heat distribution and constant air movement in the greenhouse.

In the case of *tunnel greenhouses*, if the length does not exceed 60–70 ft, end ventilation should be sufficient but in longer structures fan ventilators can be used.

FIG. 53a. Polyethylene greenhouses fitted with side ventilators. (By courtesy of J. Hanras.)

Manual ventilation has to be carried out at the right time to prevent fading of the flowers, and burning of leaves touching the film, and wilting of the plants.

The use of perforated plastics films to provide ventilation in tunnels has been extensively evaluated. It is obtained as discussed in Chapter 4, with circular holes 1·4 cm in diameter at 7 cm intervals. The minimum temperature lies between that of a closed structure and that of the outside

FIG. 53b. Ventilation of tunnel greenhouses at Lee Valley Experimental Horticulture Station.

while the maximum temperature is reduced by 5–10 °C. The use of perforated semi-forcing tunnels in France has been described by Buclon.[175]. Experiments carried out in the USA[176] have shown that the use of perforated plastics film covers gives earlier crops of cucumbers and tomatoes.

The advantages of natural air displacement are that the divergence of temperatures is reduced, humidity control is good, fertilisation of the flowers is increased and fruiting is improved, the supporting framework for ventilation is eliminated, and labour is reduced.[177]

LINING OF GREENHOUSES

Except in greenhouses which are already constructed of layflat polyethylene, lining the various structures has considerable value for the control of the temperature and the level of humidity, and thereby minimises thermal shocks (see Fig. 54).

FIG. 54. Use of polyethylene film for lining a glass-house.

Tests carried out at the National Institute of Agricultural Engineering, Silsoe, in the UK, showed that a glass-house lined with polyethylene film had 38 % less heat loss than an unlined house while the transmittance of solar energy was reduced by only 4–15 %.

Fixing

The film, usually of polyethylene, is attached in a parallel direction at about 5 cm from the exterior wall of the glass-house or GRP or PVC greenhouse, so as to isolate a blanket of air. It is fixed to the ground or to the lower part of the supports, and at the top, with twine or galvanised wire coated with an insulating emulsion which does not cause stress cracking (as discussed in Chapter 4).

Temperature
The 'air blanket' reduces the convection losses between the greenhouse and the exterior. The temperature coefficient, which is 6·2 for polyethylene sheet, falls to 2·72 for a double sheet and when the greenhouse is ventilated the coefficient changes from 11·6 to 2·4. The temperatures in the greenhouse rise only slowly and savings in heating of the order of 30% can be made.

Condensation
Because of the impermeability of polyethylene to water vapour, condensation on the interior surface of the greenhouse occurs when the surface temperature drops below the dewpoint and this has always presented a problem. The lining prevents the possibility of droplets falling on the plants, and ensures that the water collects on the upper surface of the inner film of polyethylene and then runs down between the inner film lining and the outer wall of the greenhouse.

Note: The fall of droplets from the internal wall of a single-wall greenhouse can be prevented by spraying the surface with a wetting agent which ensures that the water droplets coalesce into a continuous film which runs down under gravity. The wetting agent must have only a low tensio-active effect in order to avoid stress cracking of the film; this seems to have been overcome by 'Sunclear' which originated in the USA and is now generally available.[178]

TECHNIQUES IN THE FIELD

It has already been indicated in the Introduction that one can identify in the field the practical uses of plastics from which the major industrial and commercial activities have been derived. These latter can have an instructive aim linked to the practical techniques and this has been already done, in a way, in the treatment of the subjects considered in Chapter 3, namely: packaging, silage, water storage, sterilisation and fumigation, farm buildings, irrigation, drainage and equipment.

Mulching

Hand Laying

For small areas the growing bed is prepared with a slightly convex surface on a well raked soil with shallow trenches 10 cm deep and sides inclined at 45° to receive the film. The film is unrolled and held down at the two edges by burying them with soil. Circular holes for planting the seeds or young plants are cut with a hot perforator (as described in Chapter 4). Holes made by cuts in the shape of a cross should be avoided as these encourage tearing.

Mechanical Laying

This technique has developed rapidly with the introduction of special equipment. There are four main types:

(i) equipment which rolls out flat film over the soil, which is widely used in Europe (Fig. 55);

(ii) equipment which prepares the ground in ridges and then unrolls the film over them (used in the USA);

(iii) equipment which sows the seed and unrolls the film in one operation. A slit is made in the film at regular intervals and a seed is pushed through by a mechanical finger to the required depth. The young plant then pushes through the lips of the slit. In the case of lettuce, round holes of 7–8 cm diameter are cut at regular intervals by a heated cutter and then the seed is dropped directly onto the soil;

(iv) the type which can operate at a speed of 1–1600 kph which corresponds to one or two plants every 19 sec. The machine requires one or two people to put the plants in the holders which can cut the slits or holes in the film and do the planting at the same time.

Applications. Vegetable and fruit crops grown in the open, e.g. asparagus, aubergines, Swiss chard, celery, chicory, cauliflowers, cucumbers, gherkins, courgettes, shallots, strawberries, French beans, lettuce, melons, water melons, peppers and tomatoes (this list is given for more favourable climates than exist in the UK); non-food crops, e.g. tobacco, maize (for cattle feed) and cotton; and fruit crops, e.g. vines and fruit bushes.

Note 1: In the last category it is necessary to make a central hole greater than the section of the trunk in a square of film of about 1 m sides and a slit parallel to one of the sides to allow for planting. It is then covered with soil as are the edges of the square.

Note 2: Traditionally strawberries have been planted in two rows on the same ridge but in the American method of double ridge

118   PLASTICS IN AGRICULTURE

FIG. 55. Mechanical laying out of polyethylene mulch film.

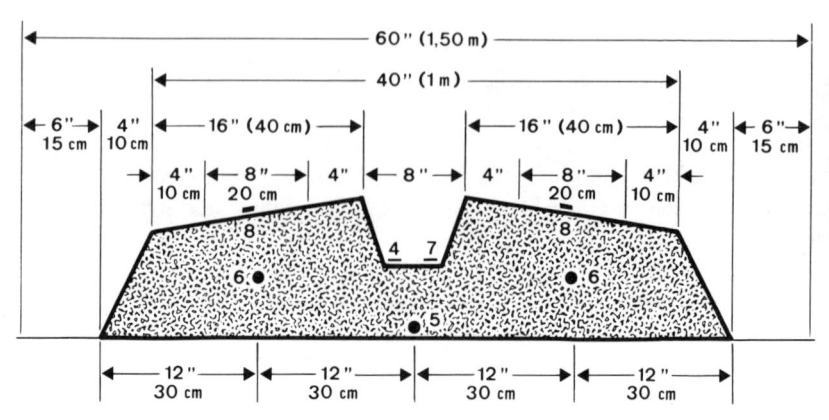

FIG. 56a. Double ridge cultivation of strawberries.

FIG. 56b. Ridge cultivation of strawberries with trickle irrigation.

cultivation a modified shape has been introduced (Fig. 56a), which increases the planting density by about 30%; irrigation is no longer carried out on the ridges but in a small central drill which has the advantage of leaving the ridge dry for the fruit crop. Figure 56a is taken from a paper by Buclon[179] and the numbers 4 to 8 indicate the location of various irrigation systems which were being assessed. It has been shown that the increase in yield with normal trickle irrigation is 9–19% and that, for the double ridge bed, the mid-furrow irrigation gives as good results as trickle irrigation on the two ridges and better results than trickle irrigation with a single tube (Fig. 56b).

Silage
Silo Towers
Such silos can be 3–4 m in diameter and height. They are formed from metallic mesh, 10 mm square, from 3·9 mm wire or from semi-rigid PVC or polyethylene lined with a plastics film of 150–200 μm thickness. A formed galvanised tube with a union at one end allows a polyethylene or PVC film to be used as a liner for concrete silos with a plastics foam intermediate

layer; this type of silo can be used for the storage of large quantities of cereals, fertilisers and harvest crops. Large silos of GRP sandwich construction have been erected on an experimental basis in Hungary; a silo of 400 m^3 was designed for green fodder and one of 800 m^3 capacity for the storage of grain.[180]

A recent publication[181] has described a cheap form of construction which could have wide application where small silos are needed in rural areas. The actual container is in the form of a large sack made of woven polyester fabric. Loops are attached to the top of the bag so that it can be suspended. The container narrows towards the bottom like a funnel and is fitted with a slide at the base for discharging the contents. The silo is used in a suspended position, being supported by a simple structure: the weight of the stored produce is carried by a steel tube attached to the sack and resting on bars. The silo is filled from the top, the contents being protected from the weather by the provision of a rigid roof.

Vacuum Silage

This was developed in New Zealand where grass was conserved and stored between two sheets of polyethylene which were sealed together at the edges and then the air contained inside was removed by means of a vacuum pump.

Basically the method is to lay out a ground sheet on which the stack of silage is built up in the normal way. A top sheet is then placed over the material and sealed to the ground sheet by means of a length of 'Strip-Seal' which acts as a plastics zip-fastener. The air from within this sealed container is then extracted by means of a suitable vacuum pump via a perforated tube sited beneath the top sheet and connected through it by means of a specially designed stack valve. The seal can be opened and closed as required without damaging the sheeting (Fig. 57).

The technique has met with only limited success in the UK due to the difficulty and tediousness of manipulating the 'Strip-Seal' and the problems of sheet damage, although this can be minimised by using heavier gauge materials.

In New Zealand, developments have taken place in the construction of permanent vacuum pads with the bottom sheet sealed into the concrete base and the top sheet, usually of heavy gauge polyethylene film, is sealed to special brass strips set into the concrete around the perimeter of the pad.[182,183]

The different types of silos for fodder storage have been considered in Chapter 3 (see Figs. 18 and 19).

Films. These should be of good quality so as to be resistant to snags and

FIG. 57. Air evacuation from silage clamp. (By courtesy of H. R. Spice.)

tears (determined by the tear strength test), and also to ageing by oxidation and UV radiation. Growers, particularly in France, are turning more and more to films which conform to the required specification, especially as regards the black film of 150 μm thickness used for grain silos and fodder silos from which the air is evacuated. A discussion on film quality and the special requirements for silage has been published by Ebel.[72]

Soil Sterilisation
Sterilisation of the soil is practised extensively in California and special equipment has been developed for treating the soil by the injection of liquid fumigants under nitrogen pressure to a depth of about 15 cm (Fig. 58). The area is covered with film from a long roll (1000 m in length), each strip being joined to the previous one by an adhesive. The final width is covered with soil to a sufficient breadth to contain the fumigants. Work reported by Hall[184] has shown that film of 38 μm thickness has advantages over thinner material because reduced dosages of fumigants can be used to produce satisfactory results. Other studies on the influence of polyethylene permeability on the effectiveness of soil fumigation on plant response and

FIG. 58. Soil fumigation equipment for injection and laying of plastics film.

weed control in strawberries have been carried out at the University of California.[185]

Water Management
Irrigation—Supply Piping
The main supply pipes, of rigid PVC or polyethylene, are buried, while the secondary pipe systems which feed the water to the actual distributors are laid on the top of the soil. Valves and flow-meters are included for regulation of the water supply. Such a watering system has a range which varies between 80 and 200 m (Fig. 59a).

Micro-tube drippers. There are many systems available and the following are some of the better known:

The *twin-wall* system consists of two sheaths of 2·5–3 cm width in black low density polyethylene of slightly different diameters (thickness 100 μm) and welded internally; the water enters by the inner tube and passes through holes to the outer tube from which it flows through holes in the outer wall. There are four outlet holes in the outer tube for each inlet hole in the inner tube. Such a system has a negligible pressure drop over 60 m and for very high outputs it is possible to extend the length to 180–200 m (Figs. 59b and c).

The *bi-wall* system is a twin-tube pipe in which the larger tube acts as the

PLASTICULTURE IN PRACTICE 123

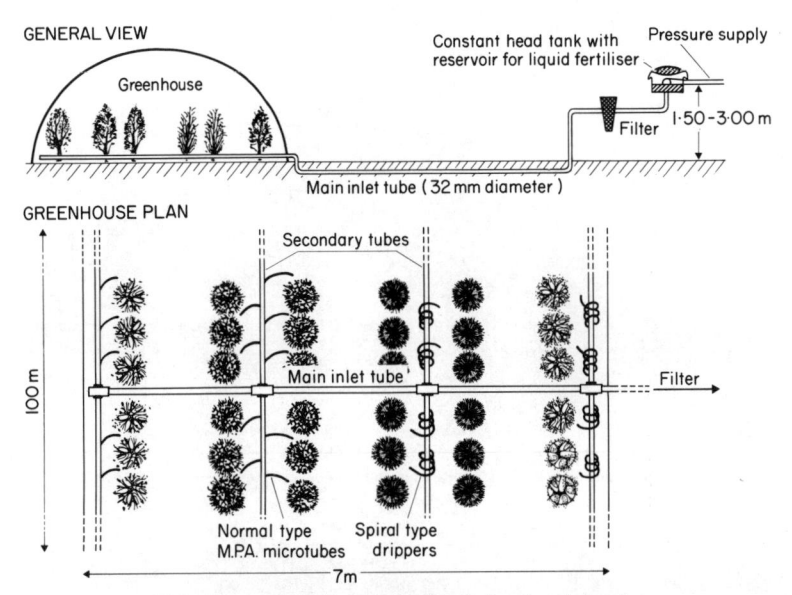

FIG. 59a. Greenhouse trickle irrigation layout.

water supply pipe to the smaller distributor tube. It is made from extruded polyethylene which is irradiated to prevent stress cracking. Various flow rates and orifice spacings are available to cater for different crops and climate conditions. The normal operating pressure is 0.3–1 kg/cm^2 (5–15 psi) and, as with all trickle and drip systems, the water must be filtered. Both this and the twin-wall system are supplied in coils which can easily be run out over the ground (Fig. 59d).

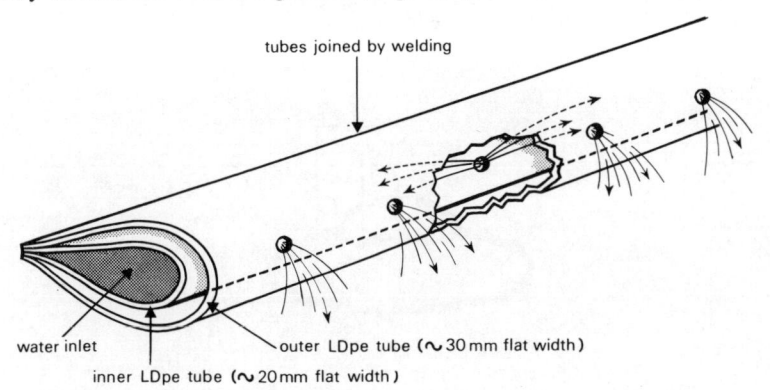

FIG. 59b. Detail of twin-wall system; LDpe = low density polyethylene.

Fig. 59c. Use of twin-wall system in Senegal with simple feed tank.

The *porous tube* system (*Viaflo*) is different from the previous systems in that it is based on a flat tube made of porous polyethylene, the porosity being of the order of 50%. The water seeps through the whole circumference of the tube so that it is wet over the complete surface. Variations of this type include reinforcement with an edging of black polyethylene; in another, one side of the tube is thicker than the other so as to provide protection against degradation.

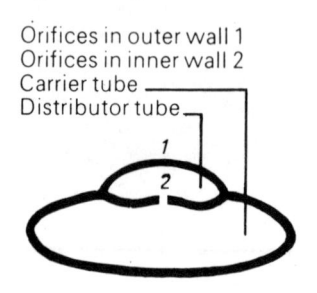

Orifices in outer wall 1
Orifices in inner wall 2
Carrier tube
Distributor tube

Bi-wall tubing cross section

Fig. 59d. Detail of bi-wall system.

There is now a multiplicity of *drip-head* systems in which the drip control fittings are attached directly to the wall supply tube. The 'Drip-Eze' drip nozzle is moulded in black polypropylene and it is claimed to provide water at a more consistent rate than other systems without risk of blockage.

The 'Key-Emitter' (Cameron) is moulded from polypropylene. It is a drip-head which is designed to ensure constant water flow in spite of any fluctuations in the pressure of the supply. It has a pressure compensating membrane which self-adjusts to ensure a regulated discharge at each outlet

FIG. 59e. Close-up of micro-tube irrigation.

point over a wide pressure range. The units are fixed onto the supply pipe into holes which are punched with a special tool and secured by patented attachment lugs.

In *micro-tubes* the dripper takes the form of a capillary tube, the amount of water emitted being determined by its length and internal diameter. The supply tube, in low density polyethylene, has an external diameter of 25 mm (internal diameter 21 mm) and the micro-tubes have an internal diameter of 0·3–0·7 mm; the water pressure is 0·8–1·5 kg/cm^2 (13–23 psi). The microtubes are simply fitted by forcing them into holes pierced in the water tube (Fig. 59e).

Perforated tubes are rigid pipes of either PVC or polyethylene, drilled with holes of 1·6–2·5 mm diameter at about 70 cm intervals; a sliding collar provided with grooves on the inner surface channels the water onto the plants. The tubes are supplied in lengths of 80–100 m and operate at a water pressure of 0·8 to 1·5 kg/cm^2 (13–23 psi).

Whatever system of irrigation is used, it is necessary to take account of: (a) the soil structure and texture, in order to determine the amounts of water needed; (b) the data on the potential evapotranspiration (ETP) and possibly the rainfall, in order to determine the frequency of watering. Inside a greenhouse, the ETP can be calculated according to the formula of de Villele (INRA Avignon–Montfavet, Laboratoire de Bioclimatologie):

$$\text{ETP, mm H}_2\text{O per day} = a \times \frac{GR}{L} - b$$

where GR is the global radiation in the greenhouse in cal/cm^2; L is the latent heat of vaporisation; and a and b are experimental coefficients from the regression equation.

As a first approximation, the correlation found in the greenhouse is independent of the climate established inside (temperature and humidity) but remains directly dependent on the solar climate of the region and the optical properties of the wall.

Water Reserves

Hill lakes and reservoirs. A black film of low density polyethylene or PVC is stretched over the whole area of an excavation in order to contain large volumes of water. After removing all the stones and anything else which might cause perforations, the films being laid out are joined either by welding or with a mastic and adhesive tape and are then anchored by burying the edges in trenches of 30 cm depth; they are then protected around the periphery of the bank at the top by means of paving slabs or

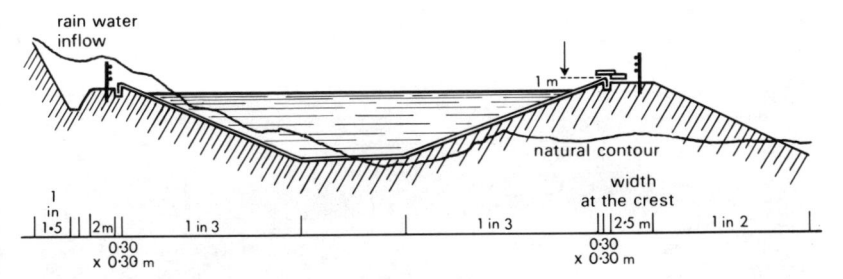

FIG. 60a. Hill lake.

grass. The sides of the water reservoir must be less than 40° with the horizontal. The film, the thickness of which is generally 200–400 µm is covered with a layer of 10 cm of sand and this is then topped with 10 cm of gravel (Figs. 60a–c). An outlet for the water can be formed by welding a plastics sleeve into the side-wall of the reservoir through which the exit pipe passes and then the joint can be made tight with a special adhesive tape. Many different modifications can be introduced.[186–188]

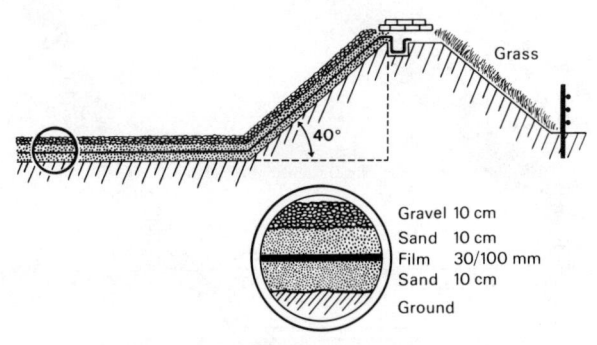

FIG. 60b. Detail of water reserve.

Low-capacity water storage reservoirs (of 2000 to 3000 gal capacity), which can be very effective in arid areas where evaporation losses are high, are constructed using black tubular polyethylene film (2·5–3 m flat width) with a thickness of 150 to 200 µm (Fig. 60d). The water is thereby contained within a cover so that evaporation losses are eliminated.

Water catchment. Where no water sources such as streams are available, small water catchment areas can be formed by lining depressions in the ground with black polyethylene and catching the rainfall during wet weather (Fig. 61).

FIG. 60c. Water reserve lined with polyethylene film.

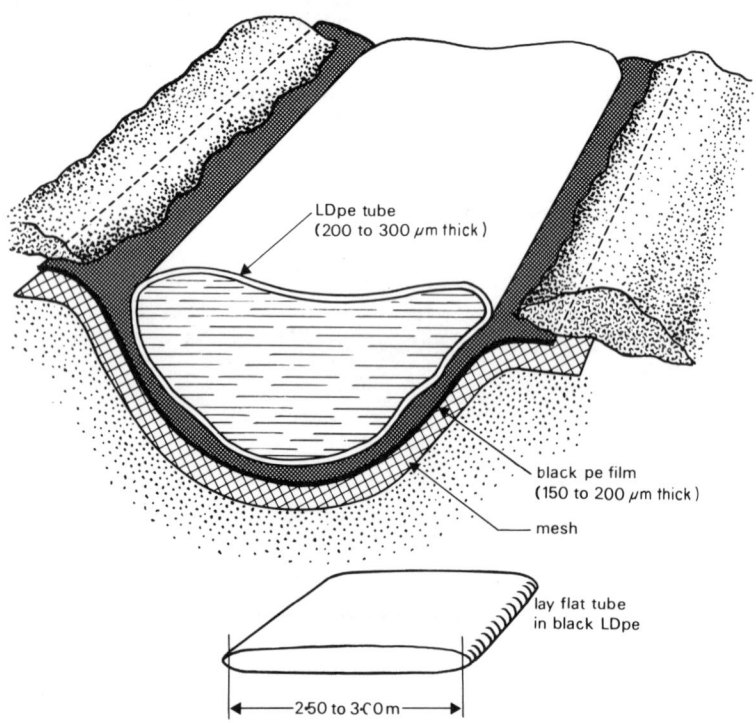

LDpe tube (200 to 300 μm thick)

black pe film (150 to 200 μm thick)

mesh

lay flat tube in black LDpe

2·50 to 3·00 m

FIG. 60d. Section of storage reservoir with tubular polyethylene film.

FIG. 61. Water catchment areas in desert regions. (By courtesy of F. Buclon.)

Catchment areas for reservoirs can be constructed by covering a large area adjacent to the storage reservoir with polyethylene film which is then covered with gravel which holds the film in place and protects it from the sun. The catchment area has a slight slope so that all the rainwater is collected and fed into the reservoir. One millimetre of rain represents 1 litre of water for every square metre of soil.[189]

Dams. Films of 200 μm thickness can be attached to posts spaced at intervals of 1–1·2 m to create a barrier when the soil is sufficiently impermeable; such dams are used in the cultivation of rice to prevent inundation of areas on slight slopes on the banks of rivers (Fig. 62a).

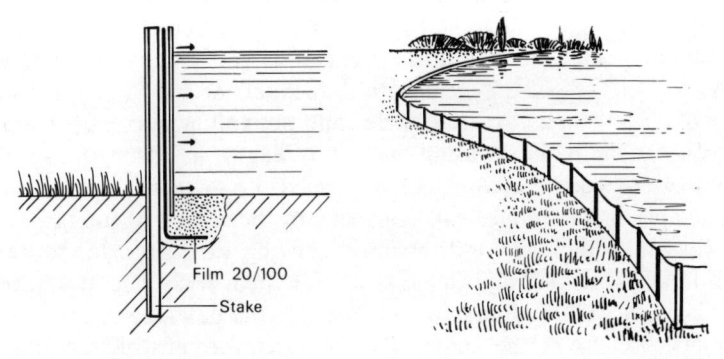

FIG. 62a. Low embankment—protection against flooding in rice culture.

The flooding problems in Venice have led to the development of a flexible structure to test the feasibility of constructing a larger dam. The model, which is formed from rubber-coated fabric, is 62 m long, 5 m high and 12 m wide (Fig. 62b) and is maintained in position by piles driven into the bottom of the sea bed; the soil is provided on the bottom by adherence of the flexible membrane to the sea bed.[190]

FIG. 62b. Experimental dam at Punta Pila, Italy.

Channels and canals. Black film is used to line water channels to prevent losses by infiltration into the surrounding ground.[191] Low density polyethylene is normally used, but high density polyethylene film is more suitable where high temperatures are encountered.[192] Large flat tubing of black low density polyethylene, with a thickness of 200 to 300 μm and a width of 2·5 to 3 m, is also used where conditions allow; it is easily handled and can be readily joined using metallic tubing.

Rubber-coated fabric, which can be formed into a channel by attachment to a simple metal frame, can conveniently be used for emergency or temporary installations, for diversion of existing waterways, and for rapid creation of water supply and distribution networks. The structure is modular and easy to transport and install; the advantages are mainly economic because of the limited time required for construction and the possibility of re-use.

Perforated layflat tubing for watering. Black low density polyethylene layflat tubing with a flat width of 6–20 cm is perforated with holes of 0·2 to 0·3 mm diameter at every 20–30 cm and is simply laid out on the soil. The efficient operating length is 40 m at a pressure of 0·5 kg/cm^2 (7·5 psi).

Note: The joining together of two sections of the larger layflat tubing is done by means of two hoops formed from 4 mm steel wire with rims which fit one inside the other; the water pressure ensures adequate tightness.

Drainage

A general study on the situation as regards irrigation and drainage in France has been published by Quentin.[73]

Drains. In Europe PVC pipe is normally used and this can be in either the slotted straight or perforated corrugated form. The corrugated pipe is available in continuous coiled lengths of up to 200 m, the diameter varying between 50 and 125 mm. Each coil is supplied complete with one straight connector which simply snap-fixes onto the next pipe. The rigid slotted pipe is supplied in lengths of 6 m, each pipe being complete with an integral socket so that it can be joined simply by a push-fit to adjacent pipes. In unstable soils it is recommended to use straight pipe which is encased in coconut fibre; this increases the diameter of the drain and reduces the risk of blockage.

The filtration area of the drain is of the order of 30 cm^2 for each 1 m run of pipe and this is achieved by maintaining a perforation size of between 0·6 and 1·7 mm which varies according to the different specifying authority. The British Standard for plastics pipe (BS 4962:1973) for light duty subsoil drainage specifies a maximum perforation width of 1·25 mm. The standard also covers perforation distribution—at least one perforation per 120° segment in every 150 mm length of pipe—this uniform distribution providing for the most efficient water intake. British Standard 4962 also requires that the pipe has adequate resistance to degradation and will not deteriorate under the conditions of use.

Drain laying. One of the main advantages of the use of plastics pipe is the minimisation of damage to the structure of the soil by the reduction in the weight of the loads to be transported across wet land (as compared with heavy clay tiles or concrete pipes). The actual laying is best carried out using equipment with wide caterpillar tracks to reduce the soil loading, the actual pressure being about a quarter of that for tyre-tracked machines. Two different types are normally used: trench cutter and layer, and moleploughs.

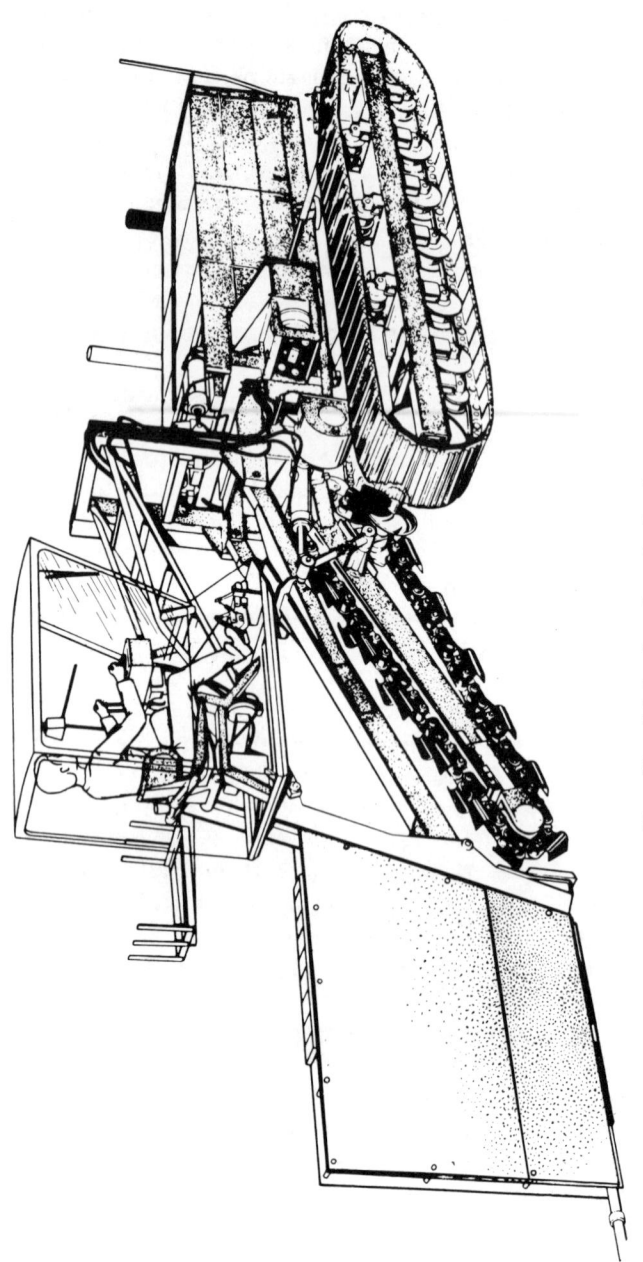

Fig. 63. Mole-plough and pipe layer.

The trenching machine (Fig. 63) is equipped with a guidance system controlled by a laser beam so that the depth of the trench can be automatically regulated to a precise slope of 0·01 %. The trench is cut to a width of 20 cm and the drainage pipe is fed into the bottom through a metallic casing to avoid subsidence. These machines allow drainage pipes to be laid at the rate of 3 km per day as compared with 10 m per day for hand laying.

Drainage using expanded plastics. Expanded polystyrene granules can be used as an aid to drainage in that they form a bacteria-proof filter and prevent the drains from silting up. The amount required varies from 1 to 3 m³ for every 100 m run of drainage. The granules can also be used as the porous back-fill for channel drainage. Trenches are cut to a depth of about 70 cm with a width of 8 cm; these are then filled with expanded polystyrene granules, the amount required being 4–5 m³ for every 100 m length of channel.[84]

Construction of Protective Tunnels
General
Although there is now a wide range of commercial structures based on plastics materials available to the grower, the principles of the construction of basic structures are still the same as set out by Spice in his book *Polythene Film in Horticulture*,[106] which was published in 1959. Many workers have studied the various factors which affect the microclimate and the performance of crops grown under such conditions. The fundamental aspects have been studied by Professor Nisen[193] and workers at the Institut National de la Recherche Agronomique at Montfavet (France).[194] In the UK the practical aspects of the construction of walk-in tunnels have been the subject of extensive work at the Lee Valley Experimental Horticulture Station, which has examined the various types of film available and made recommendations on the design of the supporting structure. The majority of UK suppliers of plastics structures have adopted these recommendations.

In France, the approach in design of greenhouses has been different and the models available commercially are much more sophisticated, particularly as regards ventilation and general quality of the structure, with a resulting increase in the cost. There is therefore some incentive for growers in France, where simple protection is all that is needed in many instances, to construct their own protective structures.

Different Types of Tunnels
There are numerous models and these can be classed in two groups: those which use reinforced PVC—these tend to be the more developed and

costly—and those which use low density polyethylene, EVA and plasticised PVC which tend to be the simpler type and the covering has to be replaced every few years.

Low Tunnels

Although the use of this type of protection requires quite a lot of manpower—for this reason its use in the USA is on the decline—low tunnels are widely used in France and to some extent for the cultivation of strawberries in the UK. The tunnels most widely used in the UK are based on the system developed at ICI Plant Protection Division Research Station at Fernhurst in the UK. It consists of six or eight s.w.g. wire hoops with eyes which are placed down the row at about 75 cm intervals and then covered with thin (38 μm) polyethylene film. Lengths of baler twine passed through the eyes on the hoop at ground level and over the film hold it securely in place. When the crop is finished, the film is removed and burnt, while the twine is retained for use the following season (Fig. 64a).

A similar system is used in France, except that the film is held in place by lengths of high tensile wire which pass from one eye on the hoop to the other. The hoops are 2·5 m long, the eyes being formed 27 cm from the ends (Fig. 64b). Film of 100 μm thickness is used so as to be strong enough for frequent manipulation. Film of 2 m width is required, although the arc

FIG. 64a. Low tunnels in UK for strawberry cultivation.

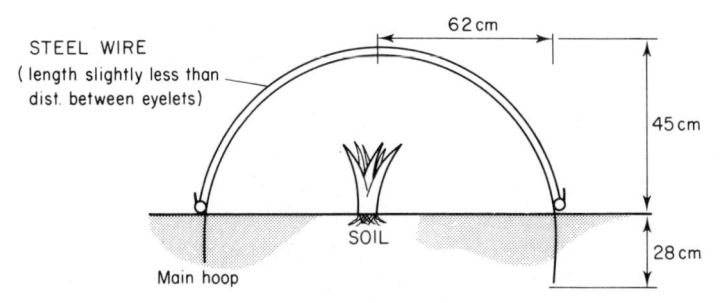

FIG. 64b. General design of 'Nantais' tunnels.

perimeter is only 1·8 m so that the edges can be turned up to form a hem, thereby reducing the possibility of edge tear and forming pockets which collect rain water and hold the film down to the ground. In order to provide ventilation, the film is slid up between the main hoop and the retaining wire so that the sides are opened up.

Construction of Plastics Greenhouses
Control of the Atmosphere
It is essential that there are means of making the most effective use of the various environmental factors, namely;

the transmission of energy waves coming from the sun and soil;
temperature arising from the same origins; and
humidity from the air, the soil and the plants under cultivation.

Air displacement and renewal must be controlled in order to help with transpiration of the plants and to avoid condensation of water droplets on the internal walls as these encourage the development of plant diseases.

Note 1: Relative humidity of the air. The water which may condense on the walls evaporates if the temperature rises by gradually absorbing heat (580 cal/g). The thermal shocks are thereby less severe than with glass, with which condensation is less likely.

Note 2: Thermal losses. The convection losses between the wall of the film and the exterior are of the same order as for glass.

Miscellaneous
Economics
The economic aspects of the use of plastics for crop cultivation have been reported by Bry and Daverat-Julian[195] who made a comparison of the costs

of various types of protective structures and also of the production of lettuces, tomatoes, melons and peppers for different regions in France as at the beginning of 1976.

Energy Saving

Double walls for greenhouses and protective structures. Reference has already been made to the value of double walled and lined structures and also to large width films which help in installation by forming a continuous surface and reducing the number of joins. The maximum width of polyethylene sheet available is of the order of 12 m. In the structure shown in Fig. 65, the air layer, 3 cm in thickness, is maintained by the use of distance pieces and by air injection between the two walls. The savings in heating can exceed 30%.

FIG. 65. Inflatable greenhouses for the cultivation of flowers in San Diego, California.

Reflector screens. A film aluminised on the two faces prevents losses of infra-red radiation if it is set in place during the night. Such screens help the radiation in the area at the base of the greenhouse, where plant growth is least favoured, and stimulates the reheating.

Thermal spar system. In this system (developed in Germany) polyethylene film tubing is installed in the roof section, and when the tubes are inflated the growing area is hermetically sealed off (Fig. 66). Insulation is effected by two thicknesses of polyethylene and the volume of air included reduces the heat losses by 35%. The tubes are normally inflated during the night and the air is let out during the day, when they hang limply from the fitments, so as not to reduce the light coming in through the roof.[196]

FIG. 66. Thermal spar system for roof insulation.

Expanded polystyrene insulation. Since insulation tiles are impermeable to light, their use is limited to those areas where the loss of light is minimal (up to bench height, or for covering north-facing gables) as well as for insulation of the foundations. Light reflection partly compensates for the losses. The thickness of the material used depends on the insulation required. The sheets of expanded polystyrene can be made up as portable folding screens for ease of handling.

Hydroponic Cultivation or Soil-less Culture

The development of a seed depends on three factors: air, light and the soil which supports it. To these must be added heat and nutritive elements (water, fertiliser). With a view to separating the various factors involved in development, studies on plant biology have aimed at isolating the role of the soil by using inert media such as expanded polystyrene, and polyethylene pellets in pots (such as the Plantube) which are themselves arranged in channels through which the nutrient solution flows intermittently.

In Kuwait there is a pilot scheme at the 'Center for Protected Vegetable Production' for growing vegetables, fruit and flowers and also for the production of milk and eggs. The plant was put into operation in 1952.

Hydroponic cultivation is used over an area of 200 000 m^2 on sand held in cement beds and saturated with a solution of nutrients in soft water, obtained by desalination of sea water. This is recycled through the sand bed and the nutrient content is made up as necessary.

The nutrients are obtained from household rubbish from Kuwait City and from urea produced locally (2 tons/day). The plastics greenhouses used are in the form of low tunnels so as to be less affected by sandstorms. The cover is moved by hand for ventilation or shading using, if need be, a coating of gypsum dust, which adheres to the polyethylene but can be washed off in winter.

The water is distributed at very low pressure by a channel made up of a woven band along its length and resting on the soil. The solution flows through the holes of the material (drip irrigation).

The temperature is lowered by 5°C in summer as compared with the outside by using a wet brushwood screen which cools the air as it passes through into the greenhouse.

Dairy cattle bred from cattle imported from Jersey provide the milk which is sold locally. It is intended to extend the air-conditioned buildings.

Overall, the production has reached an interesting stage but at a fairly high cost.

In the UK, a simplified form of hydroponic culture has been developed at the Glasshouse Crops Research Institute.[197-199] The nutrient solution (at 25°C) runs along a plastics gully which is set on a slight slope, the depth of solution on the bottom being maintained at less than 1 mm. The young plants are located in the gully with the roots in contact with the solution. The early trials were carried out using a gully formed in a triangular shape from thick black polyethylene film, the apex being closed to prevent the ingress of light and the growth of algae. The roots of the plants develop quickly and form a thick mat which completely covers the bottom of the gully. The solution is circulated by means of a pump and plastics pipe system which includes a make-up tank where adjustments can be made to the composition of the solution. Pre-packed mixtures of the required nutrients, containing trace elements, are available and it is necessary only to add a prescribed amount when the conductivity of the solution falls to 2 milliohms; the pH is maintained at between 6 and 7 by the addition of phosphoric acid. This system is already widely used for the cultivation of tomatoes and cucumbers. It is of interest to note that wide variations in the composition of the solution can be tolerated by the plants. Commercially produced equipment is now available for the application of this process (Fig. 67).

Cuttings, Grafts and Layering

Low density polyethylene film (thickness 50 μm) is very effective as a support and thermal regulator for the development of cuttings. Bands of

FIG. 67. Tomato cultivation using the nutrient film technique (NFT). (By courtesy of Soil-less Cultivation Systems Ltd.)

flexible PVC with a width of 1 cm are suitable for grafts; they are impregnated with small amounts of lanolin and naphthalene acetic acid. Vinyl emulsions used as paint give good protection against air, rain and micro-organisms.

Protection of Plants Against Rodents and Rabbits
This can be done very effectively by using a section of rigid PVC tubing which has a slot along the whole of its length and is slipped around the plant.

Prevention of Soil Erosion
In areas of light sand and peat soils, erosion of newly seeded plots and damage to young plants because of high winds can be a serious problem during the early part of the year. It has been established that this can be overcome by spraying the ground after planting with a dilute dispersion of a plasticised polyvinyl acetate which forms a bond and stabilises the top layer of the soil.[200] Penetration can be up to 6 mm in sandy soil when a diluted latex is applied as a fine spray at the rate of 100 gal per acre (1138 litres/ha). At soil temperatures greater than 3 °C, the latex dries to give a strong film within half an hour of the application. The economics depend on the frequency of high winds and the cost of re-drilling after the loss of the seed or young plants.

Water Savings
The optical properties of a protective structure covering, whatever the material, play a very important role in the consumption of water for plant cultivation inside the greenhouse. The method of irrigation employed also has a significant effect on water consumption. Trickle or drip irrigation, for example, can produce a saving in water of the order of 30 %. To conserve water, drainage water can be used for irrigation as long as the pesticide content is low.

Plastics Waste and Ecology
The problem lies within the general area of pollution affecting the atmosphere, soil and water. Plastics are often considered as major pollutants but fortunately it will be seen that their role in this domain is only very modest.

The problem has been put into perspective following innumerable conferences over recent years and particularly since the oil crisis in 1973 when the price of plastics rose considerably.

There is at the present time no specific problem arising from plastics rubbish. Staudinger[201] rightly distinguishes between household rubbish and litter; the latter is a particular nuisance and difficult to collect. Only about 3% of household rubbish is plastics and in France the total amount for the large towns is about a million tons a year, of which about 40% is derived from packaging (paper and cartons 20%, wood and crates 5%, glass 5%, tins 4%, various 3% and plastics 3% (20 000 tons)).

Plastics are very apparent as rubbish because of their low density rather than their mass (density 0·92 to 0·96 for the polyethylenes, and 1·4 for PVC), but they are easily compacted and it is possible to envisage collection and re-use. The main types of plastics which appear as rubbish are limited mainly to the thermoplastics polyethylene, polystyrene and PVC. In the UK in 1969 the breakdown of plastics materials used in packaging was low density polyethylene 57%, high density polyethylene 12·2%, polystyrene 16·2%, PVC 3·7% and others 11%.

Waste Problem for Growers

It is necessary to replace the film covering of greenhouses from time to time and the problem arises as to what should be done with the waste material. Polyethylene burns readily without creating any obnoxious fumes except for smoke; PVC is more difficult because hydrogen chloride is given off when the material is incinerated and this can be both obnoxious and hazardous in high concentrations. The general problem of indestructibility has been overcome, particularly in the case of mulch, by the introduction of photo-degradable grades of polyethylene which embrittle after a prescribed period and then rapidly disintegrate.

In Japan, disposal methods currently being used are regeneration and incineration. Attempts have been made to reprocess waste film by recycling for lower grade products such as shoe soles, and the National Agricultural Association operates its own regeneration plant. The main difficulty lies in the coordination of waste collection and in ensuring an uninterrupted supply for the reprocessing plant. The problem has been overcome to some extent, not only by the introduction of photo-degradable materials, but also by the improved life of plastics whether they are used in the form of film for greenhouse covering or for crates for crop collection and dispatch. For the average grower the weight of waste plastics materials during 1 year is relatively low and this can be incinerated without too much difficulty if it is done sensibly when disposing of old wood and other easily burnable materials.

Chapter 6

Standards and Specifications

GENERAL

The development of the use of plastics in agriculture has been most actively concerned with improved crop production and it is not surprising that efforts have been concentrated on establishing standards for the polyethylene films which are the most widely used. In the UK, this has been done by the British Agricultural and Horticultural Plastics Association and the standard[102] which was published in 1976 is based on that which has been drawn up in France by the Centre d'Étude des Matières Plastiques (CEMP)[101] although there are some differences.

There also exist standards for polyethylene[202-205] and PVC[206-209] pipes; although they have not been drawn up specifically for agricultural use they are very relevant to all the applications where pipe systems are involved, particularly for irrigation where a high pressure water supply is used.

Mention has already been made of the standards which exist for drainage pipes.

STANDARDS TESTS

Objectives
The aim of the numerous tests is to provide data on certain physical properties which will enable the manufacturer to control the quality of his product and assure the user that it will meet the specified conditions of service. In the French standard for polyethylene film[101] some differentiation is made between different quality films which will give different performances in the field and which naturally will differ in price.

Standards Tests for Films

The standards tests which are set out in the French and British standards for polyethylene films are similar although the French one includes black films for mulching and differentiates between two films grades, one-star and two-star, which have different lives when used for covering protective structures. Agricultural films manufactured to these standards are generally available and it is useful to specify the tests and the limits of the results which are defined by the specification.

Taking of Samples

This is done, according to the dimensions laid down, both in the direction of the length and the width of the roll, which can be any length and up to about 11 cm wide.

Mechanical Properties

The following properties are measured:

1. the uniformity of the thickness within the limits laid down;
2. the elongation in a transverse direction; this must be less than 3·5% and the shrinkage in the longitudinal direction is less than 2% after heating for 20 min in boiling water;
3. the yield stress; this must be greater than 0·95 da N/m^2 (or 0·97 kgf/mm^2) in both directions for a one-star film and the maximum tensile stress is 1·62 da N/mm^2 in both directions for a two-star film;
4. the elongation at break; this must be 350% in both directions for a one-star film and 450% for a two-star film;
5. artificially accelerated ageing must show a 50% reduction in the elongation at break after 8 days' exposure for a one-star film and after 66 days for a two-star film when the thickness is 200 μm;
6. the carbon black content and its uniform dispersion for black mulch films; these are critical to ensure adequate life in use.

Marking of Agricultural Films according to the French Standard

The monogram of the Standard accompanied by one or two stars must be printed continuously at regular intervals on the film with ink or by hot stamping (Fig. 68).

Note: A special grade is now available for black films for silage use.

Supervision and Control

The supervision of the manufacturing standard is carried out by the processor at the factory, who must maintain a record of the results obtained.

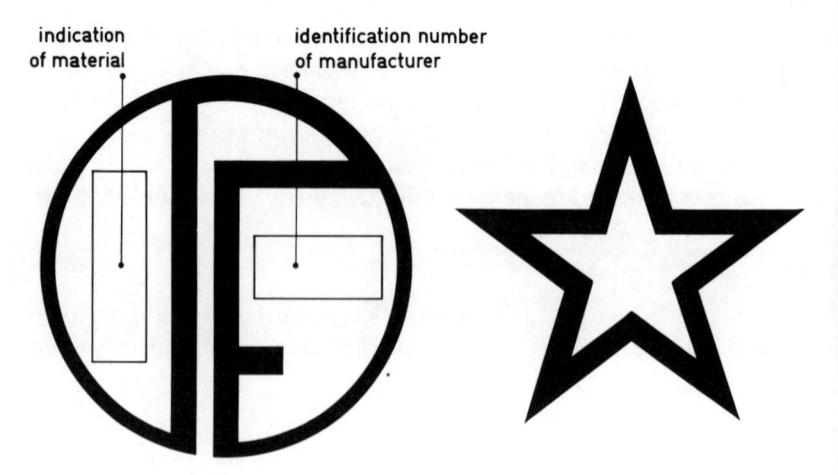

FIG. 68. CEMP marks for films for agricultural use.

TABLE 15
SPECIFICATION FOR TRANSPARENT FILMS

Property	Measurement on the film One-star	Measurement on the film Two-star	Measurement on the polymer One-star	Measurement on the polymer Two-star
Density (g/cm^3)			>0·917	0·917–0·922
Melt flow index			<2	<1
Uniformity of thickness	The thickness must not vary by ±15%			
Yield stress (kgf/cm^2)	>97 in both directions	—	>98	>110
Tensile strength at break (kgf/cm^2)	—	>165 in both directions	>130	>150
Elongation at break (%)	>350 in both directions	>450 in both directions	>375	>450
Stress crack resistance			>250 h	>500 h
Laboratory ageing resistance for film of 200 μm max. thickness	>8 days	>14 days		

TABLE 16
SPECIFICATION FOR BLACK FILMS

Property	Measurement on the film		Measurement on the polymer	
	One-star	Two-star	One-star	Two-star
Density (g/cm^3)			>0·930	>0·930
Melt flow index			<2	<1
Uniformity of thickness	The thickness must not vary by ±15%			
Yield stress (kgf/cm^2)	>92 in both directions	—	>98	>112
Tensile strength at break (kgf/cm^2)	—	>194 in both directions	>130	>165
Elongation at break (%)	>300 in both directions	>400, longitudinal >570, transverse	>375	>600
Stress crack resistance			>250 h	>250 h
Dart test strength	—	>210 g		
Extension or shrinkage at 100 °C	Transverse extension <3·5% Longitudinal shrinkage <2%	Longitudinal shrinkage <5%		
Carbon black content (%)	>2·30	>2·30	>2·30	>2·30
Dispersion of carbon black	Equal to or better than the datum point, 1·5		Equal to or better than the datum point, 1·5	

These results are checked from time to time by the Ingénieurs-Contrôleurs of the CEMP who also take samples; these are tested at the Laboratoire National d'Essais du Conservatoire and the reports are forwarded to the Standards Committee which includes members of the polymer producers, the film extruders and official bodies. This committee considers admissions for entry and makes observations for correcting manufacturing faults and withdrawing manufacturing permits.

The standards for agricultural films according to the CEMP (No. 37/3 June, 1974) are listed in Tables 15 and 16.

BAHPA Standard 'Polyethylene Film for Greenhouse Covering'
The standard drawn up by the British Agricultural and Horticultural Plastics Association is rather more limited in its scope and relates to

polyethylene film suitable for two seasons as a greenhouse covering, i.e. the equivalent of the CEMP two-star grade.

Properties of the Polymer
Density: Specimen density of the polymer shall be within the limits 917–928 kg/m^3 (0·917–0·928 g/cm^3) as determined by BS 2782, Method 509B.
Melt flow index of the polymer shall not be greater than 1·0 as determined by BS 2782, Method 105C.

Mechanical Properties of the Film
The *tensile strength* of the film in each direction shall be not less than 140 kgf/cm^2 as determined by ASTM D 882-67 Method A.
Elongation at break of the film in each direction as determined by ASTM D 882-67 Method A shall not be less than:

80–124 μm thickness	350%
125 μm thickness and above	400%

Elongation and shrinkage at 100 °C are measured by the CEMP 37/3 standard method and the same limits are adopted:

in the longitudinal direction	2%
in the transverse direction	3·5%

Edge fold strength of the film shall show below 50% failure using 10 specimens per fold, at the following dart weights

80–124 μm thickness, dart weight	90 g
125–149 μm thickness, dart weight	165 g
150 (or above) μm thickness, dart weight	190 g

Outdoor durability: The film is designated as 'two season' quality, i.e. it shall perform satisfactorily as a greenhouse covering under UK climatic conditions when used to clad tunnels and multi-span greenhouses based on the Lee Valley design. An accelerated exposure test which has proved acceptable for forecasting the outdoor durability utilises a 'Weather-o-meter' in which specimens cut from the sheet are exposed to UV radiation.
The film of the specified thickness shall retain at least 50% of its original elongation after exposure for 1000 h under the specified conditions. This method differs from that used in the CEMP standard in which a special

Material	Inflammability[b]	Speed of combustion[c]	Amount of smoke[d]	Colour of the flame[e]	Temperature (°C) of melt[f]	Reaction of fumes to litmus by heating in a test-tube[g]
High density polyethylene (d = 0·94)	2	2	0; odour of burning candle	W 0	137	N
Low density polyethylene (d = 0·92)	2	2	0; odour as above	W 0	110	N
Polypropylene	2	2	0; odour as above	W 0	167	N
PVC (rigid)	0	1	3; sharp odour of hydrogen chloride[h]	Br	120 (S)	A
EVA	1	2	2	Bl	—	A
Polyesters (GRP)	2	2	3	Bl	I	N
PMMA	3	2	0; sweetish odour	Y	125 (S)	N
Polystyrene	2	2	3; sweetish odour	Y	100 (S)	N

[a] It is not possible to identify the materials with certainty without separation of the constituents and IR spectroscopy or chromatography.
[b] 0 = non-flammable, 1 = difficult, 2 = easy, 3 = burns readily.
[c] 0 = Incombustible, 1 = Flame to keep alight, 2 = Burns slowly out of flame.
[d] 0 = None, 1 = Little, 2 = Lot, 3 = Excessive.
[e] Br = Brown, Bl = Black, W = White, Y = Yellow-green, 0 = No ash.
[f] S = softening, I = infusible.
[g] A = acid, N = neutral.
[h] *Examination for chlorine*: Heat a copper wire in Bunsen flame before bringing it into contact with the sample then reheat—an intense green coloration indicates chlorine (PVC).

ageing cabinet is used, whereas the 'Weather-o-meter' is a commercial piece of equipment which is available in many testing laboratories.

Although there are these differences between the two standards, from a practical standpoint the end performances of the films for use as greenhouse coverings are comparable.

Standards Tests for Pipes
There are many specifications and standards for plastics pipes, which are drawn up to meet a whole range of applications. The properties which are specified include: tensile strength of the material; heat reversion; hydraulic pressure resistance; and impact strength (for rigid PVC pipe).

The specification for polythene pipe (Type 50) for cold water services (BS 3284:1967)[205] applies to black polyethylene pipe and can therefore be relevant when this material is used for piping water supplies under pressure, for example in irrigation systems. Pipes are classified by maximum sustained working pressures as follows:

Class B: 6·1 kgf/cm^2, 200 ft head,   86·7 lbf/in^2
Class C: 9·1 kgf/cm^2, 300 ft head, 130 lbf/in^2
Class D: 12·2 kgf/cm^2, 400 ft head, 173 lbf/in^2

These maximum working pressures are based on water at a temperature of 20 °C and if the pipes are used at higher temperatures then the working pressures are reduced.

The specification requires that the polyethylene is pigmented with 2·5 % carbon black; this ensures that degradation by UV radiation is minimised and that the pipe is suitable for outside applications.

There are corresponding specifications for the fittings which are available for the assembly of pipe systems.

IDENTIFICATION OF PLASTICS

The identification of the various plastics materials which are used in agriculture can easily be done by technicians in the industry but there are occasions when the user needs to know the composition of the materials involved and it is useful to have available some simple tests which do not require advanced scientific equipment. The tests which are set out in Table 17 are intended to provide a simple basis for the identification of the plastics which are likely to be encountered.

Chapter 7

Results Achieved with Plastics

CULTIVATION TECHNIQUES

The various applications of plastics to cultivation techniques have been described and the benefits which accrue to the grower will vary considerably according to the climate in which he operates, to the crops being grown and particularly to the varieties which have been specially developed for the modified techniques. In the UK the weather conditions vary dramatically from one year to another whereas in more southerly climates less year-to-year variations are encountered; the British grower will therefore expect more variable benefits in the application of plastics.

Mulching using Polyethylene
The wide range of plants and crops which benefit from the use of plastics mulch have been outlined in Chapter 5. The mulch has an effect on various important features such as the temperature, the humidity, the soil structure, the presence or absence of weeds, the levels of nitrogen and carbon dioxide and the root system.

Clear transparent films favour higher soil temperatures and earlier crops but do not cause appreciable increases in actual yields. Black films are used particularly when weed suppression is required. The general results can be summarised as follows:

increase in yields of 25–100%;
the crop is sometimes 2 weeks earlier when using clear transparent films and several days earlier with black film;
elimination of certain tasks such as hoeing and a corresponding reduction in labour,
after transplanting fewer failures as compared with ordinary ground planting;

rate of cropping increases about 30% for the strawberry variety Regina and 50% for the variety Madame Houdot; helps against attacks of *Botrytis*.

Most of the development work on the mulching of crops has been carried out in the USA and southern France, and these general conclusions apply to those countries with comparable climates. Since the climatic conditions differ appreciably, it is unlikely that the same increases in yields are to be obtained in the UK. However, trials on strawberries have indicated that the use of black polyethylene of 38 μm thickness usefully advances fruit maturity and produces heavier crops.

Note on growth retardation: The opposite effect to earlier cropping, i.e. later than normal, can be obtained, for example, by covering dessert grapes in order to delay ripening. A similar effect can be produced with the flowering of roses.[210]

Low Tunnels Using Low Density Polyethylene, EVA and PVC Films

With this technique must also be associated the use of perforated or slit film which is used directly on plants without the use of supporting hoops.

Crops extensively grown under low tunnels include melons, strawberries, lettuce, carrots, and peppers, but in the UK, their use is largely restricted to strawberries and beans.

Earliness of crop: from 3 weeks to a month as compared to cultivation in the open.

Yields: higher than for the same crop grown in glass frames.

Protection against frost and inclement weather.

Economics replacement of used film is easily carried out.

In conclusion it can be stated that the investment is higher but low tunnels give better yields than mulching alone.

Mulching within Tunnels

Some aspects of the effect of mulch used in conjunction with low tunnels have been considered by Damagnez *et al*.[211]

Greenhouses Covered with Tubular Polyethylene Film

This particular type of structure was one of the earliest commercial plastics greenhouses produced in France and has now largely been superseded by tunnels covered with large width polyethylene film with a thickness of about 125 to 150 μm. The results which have been recorded for the earlier type are still relevant.

Unheated Greenhouses

Unheated greenhouses covered with tubular polyethylene film have considerable advantages over glass cold frames for melons. The commencement of gathering is advanced by 1 week and at mid-harvest, the yield is higher by a factor of two. By comparison with cultivation in the open, tomatoes are earlier by 3 weeks and the yield is tripled.

Results obtained with the cultivation of lettuce and tomatoes in tunnels covered with a range of different types of films have been reported by Brun and Laberche.[212] It appeared that, as with the melon, cultivation of tomatoes in spring is very dependent on the weather during March and April. Films of PVC and EVA are more favourable to the production of early crops, as the temperatures are low at this time of year.

Heated Greenhouses

Trials have been carried out on tomatoes in the region of Marseilles in France on tomatoes which were sown on the 7th November and then planted in a plastics greenhouse at the end of January. The first fruit were picked on 12th April. The yield by 15th May had reached 5 kg/m^2 and it had increased to 10 kg by the middle of June.

Greenhouses Covered with Reinforced PVC (Tunnels and Multi-Span Structures)

These give results which are much in line with those obtained for the same investment from a greenhouse structure constructed from tubular polyethylene film.

Structures using films based on PVC have thermal advantages although the development of IR polyethylene has minimised the difference. The advantage of polyethylene lies in the availability of film of wider and wider dimensions. Reinforced PVC has a longer life and PVC films are still used extensively in Japan.

Greenhouses Glazed with Rigid Sheets (GRP)

The high luminosity obtained with this type of greenhouse explains their success in the horticultural field, and the high quality of the flowers, particularly carnations, which can be grown in them.

PLASTICS IN TROPICAL HORTICULTURE

The use of plastics can, in many tropical areas, be very effective in improving the quality of the produce and in making better use of the natural resources. One paper which is typical of the development in this area is that of Bugalho Semedo.[110]

In the mid-1960s, the Environmental Research Laboratory at the University of Arizona began extensive research into the development of food systems for desert regions of the world. Vegetables are now being produced in many areas, where once hardly anything grew, due to the use of plastics for applications such as:

1. lining of reservoirs,
2. catchments in harvesting rainfall,
3. mulch to control evaporation and to modify the soil environment,
4. water supply pipes,
5. trickle or drip irrigation,
6. environmental control, e.g. greenhouses, low tunnels.

This was discussed by Jensen[213] at the 6th International Colloquium held in Argentina in 1974.

The severe drought conditions experienced over recent years in Africa have caused grave concern and the United Nations Agencies have concentrated on the introduction of techniques to improve agriculture; this has involved conservation and more effective use of water which requires the use of plastics pipe and film.

CONSERVATION OF PRODUCE

The improvement in the conservation and packaging of produce has been set out in Chapter 3.

WATER MANAGEMENT

The plastics materials polyethylene and PVC have for the main part replaced conventional materials in the field of pipes for potable water, irrigation and drainage (Chapter 3), and in the construction of water reservoirs (Chapter 5).

ECONOMICS

Plastics have enabled a reduction to be made in labour requirements by their contribution to agricultural techniques and the construction of agricultural buildings (Chapter 3). They have also contributed to a considerable increase in crop production which is parallel to the development of the tonnage used in this particular area of application.

THE WORLD-WIDE DEVELOPMENT OF PLASTICULTURE

The use of plastics in agriculture and horticulture was introduced into France in 1958 because in the south, where there is extensive agricultural production, wide temperature variations between day and night are encountered, rainfall is erratic and strong winds are not an infrequent occurrence. Plastics film enables plants to be protected, water to be used more effectively and improved crops to be grown. Many of the developments which have become universally adopted throughout the world were originally worked out in France and there are many organisations which continue to seek further improvements of cultural techniques and plastics often have a very important role to play in these improvements.

There are now, in many countries, organisations which have been formed to encourage and guide the more effective use of plastics materials in agricultural and horticultural applications. In the UK, the British Agricultural and Horticultural Plastics Association has been working to stimulate the use of plastics and to provide standards to enable users to obtain products especially suited to their requirements.

An international organisation (Comité International des Plastiques en Agriculture) has been formed to co-ordinate the activities of the various national bodies, and it is responsible for the planning of an International Colloquium which is held every 2 years. This always attracts delegates from many countries and has enabled cultural developments using plastics to be discussed and adopted in those parts of the world where climatic conditions present problems for the most efficient growing of produce. The international organisation CIPA is now officially designated a consultative body by the United Nations Industrial Development Organisation.

CONCLUSIONS

Plasticulture leads to a more profound, rational application of the facts of plant biology. The uses to which plastics technology can be applied, when arranged in increasing order of importance, are mulching, low tunnels, protective structures and greenhouses. The outstanding success of plasticulture in comparison with traditional agriculture, from the point of view of earlier crop production, safeguarding the harvest and making economies, particularly in the consumption of energy and water, is thus clear.

By its contribution, both to the quality and to the quantity of crop production, plasticulture is helping in the relief of hunger in the world. Unprotected cultivation is thus losing its importance in favour of protected cultivation because of the need for safeguarding crops against inclement weather conditions and growing better quality produce.

Chapter 8
Prospects for Plasticulture

GENERAL SITUATION

In spite of the fact that plastics materials are oil-based and have in consequence been subject to dramatic price increases since the oil crisis in 1973, the increase in the use of plastics for agricultural and horticultural applications continues at a rate which is generally higher than the overall growth rate for the plastics industry. This is due in no small part to the demand for vegetable and fruit crops at all times of the year; 10 years ago it was considered normal to be able to buy certain fruits and vegetables only when they were 'in season', but now improved transport has enabled produce to be distributed from the point of cultivation to all parts of the world quickly and cheaply. This has enabled those countries with favourable climates to become producing centres and in order to ensure the maximum production of high quality produce, the growers have adopted the latest techniques, many of which require the usage of plastics in some form.

USAGE OF PLASTICS IN VARIOUS COUNTRIES

There are certain authors who distinguish between 'real plasticulture'—the use of plastics for strictly agronomic uses—and plasticulture in its widest sense, which includes the commercial use of plastics for packing not only of produce but also of fertilisers and seeds and, in addition, that used in machinery and farm buildings. For this reason overall statistics must be examined with some care and useful conclusions can only be drawn from statistics which are sub-divided into the various main fields of application which have been discussed.

Table 18 shows the overall tonnage of plastics used in a number of

TABLE 18
PLASTICS USAGE IN VARIOUS COUNTRIES FOR CULTIVATION TECHNIQUES

Country	Tonnage 1973	1976	Population (millions)	Usage (1976) per inhabitant (kg)
Japan	197 000	233 700	112·77	2·07
USA	95 000	96 000	215·12	0·44
Italy	71 000	70 000	56·19	1·24
France	57 000	68 000	52·97	1·28
UK	32 000	47 600	55·93	0·85
Spain	25 000	29 000	35·97	0·80
Hungary	12 600	16 000	10·32	1·55
Argentina	6 000	5 500	25·72	0·21
Portugal	600	1 170	9·45	0·12

countries for the cultivation techniques which include mulch, low tunnels, greenhouses, silage, crop protection, water reserves, irrigation and drainage; figures are given for 1973 and 1976. The outstanding feature is the enormous usage in Japan which outstrips all other countries not only in the total amount of material consumed but also in the amount per head of population.

It is interesting to look at the breakdown of the usage in the various countries and figures are given in Table 19 for the three main applications, mulching, low tunnels and greenhouse structures.

There are several interesting features which arise from these figures. Mulching is particularly developed in those countries where water needs to be conserved, when it is necessary to heat the soil lightly in order to obtain growth, when there are many weeds and also when the country is highly mechanised and short of labour.

TABLE 19
BREAKDOWN OF TONNAGE FOR 1976

Country	Mulch	Low tunnel	Greenhouse	Other uses
Japan	25 200	27 900	45 800	134 800 (58%)
USA	18 000	4 000	10 000	64 000 (67%)
Italy	2 500	8 500	45 000	14 000 (20%)
France	9 000	7 000	5 000	47 000 (69%)
UK	100	700	1 800	45 000 (95%)
Spain	3 000	1 800	10 250	14 000 (48%)
Hungary	—	1 000	5 000	—
Portugal	117	30	1 000	23 (2%)

In countries such as Italy and Portugal practically the whole usage of plastics is as a direct growing aid in the form of film; this is the opposite for the UK where the usage for protected cultivation forms only about 5% of the total. It is surprising that the 'other uses' category for France absorbs 69% of the total when the development for mulch and semi-forcing is practised so widely: figures are available for the other sectors for 1976 and these are given in Table 20 together with the corresponding values for 1973.

TABLE 20
USAGE FOR FRANCE FOR CULTURAL APPLICATIONS

Use	Usage (tons) 1973	1976	% Increase
Mulch	7 500	9 000	20
Semi-forcing	6 500	7 000	8
Greenhouses	4 000	5 000	25
Silage / Crop protection / Water reserves	24 000	26 500	10
Drainage	5 000	7 500	50
Irrigation	4 000	5 000	25
Pots and containers	4 000	6 000	50
Nets	2 000	2 000	Nil
TOTAL	57 000	68 000	18·5%

The large usage for silage, etc., comes within film so that this accounts for over 70% of the total; this is likely to fall to some extent in view of the growth of plastics drainage piping and the wider use of trickle irrigation systems. The film usage is about 95% polyethylene with the other 5% being in PVC; this is the normal breakdown for most countries, except in Japan, where there is a large usage of special PVC films, but even in this country there is more growth in the use of polyethylene at the expense of PVC.

Usage of Plastics in the UK
The general pattern of usage in the UK is rather different from those countries where the climate is more favourable to crop cultivation; polyethylene film for mulching, low tunnels and walk-in tunnels forms only 5% of the total. The breakdown in Table 21 is given for 1976.

An interesting feature is the usage of rigid PVC sheet for crop protection as this application is mainly limited to the UK. The amount of polypropylene used for baler twine accounts for about 60% of the total

Table 21
BREAKDOWN OF THE UK MARKET FOR USAGE OF PLASTICS IN AGRICULTURE (1976)[a]

Use	Consumption (tons)
Greenhouses	
Commercial film houses	1 500 (almost entirely polyethylene)
Commercial, domestic and other greenhouses, frames and growing structures	500 (almost entirely rigid PVC)
Mulch	
Commercial usage	100 (entirely polyethylene film)
Cloches/low tunnels	
Commercial and some domestic	500 (entirely polyethylene film)
Domestic cloches and other market gardening/domestic use	500 (mainly rigid PVC)
Silage	
Commercial film covers/tarpaulins	3 500 (polyethylene film)
Commercial film covers/tarpaulins	4 000 (PVC film)
Baler/binder twine	4 000 (polypropylene)
Silos	200 (GRP/PVC)
Irrigation	
Commercial irrigation pipe	3 000 (polyethylene)
Commercial pipe and domestic hose	4 000 (PVC)
Drainage	
Commercial field drainage pipe	1 000 (high density polyethylene)
	2 000 (PVC)
Pots/seed trays	
Commercial/domestic	9 000 (polystyrene)
	1 000 (low density polyethylene)
	200 (others)
Produce boxes	
Commercial	2 500 (mainly polypropylene with some high density polyethylene)
Building	
Fences, windbreaks	4 000 (GRP, PVC)
Tools/equipment/machinery	2 000 (high density polyethylene)
Tanks/cisterns/brackets	4 000 (polypropylene)

[a] By courtesy of Imperial Chemical Industries Ltd (Plastics Division).

market for this product; this is in contrast with other European countries where sisal still dominates the scene.

EXTENSION OF PRESENT USES

It has already been indicated that the progression in the rate of usage of plastics for agricultural and horticultural applications is greater than the general growth rate for the industry. This is likely to remain the case for some time as there is continuing effort to improve food production and reduce costs and many of the techniques involved rely on the use of plastics materials.

Many of these cultural techniques are now being adopted by countries in the developing world, particularly in those areas where there is a shortage of water, and as these applications are, for the most part, still in the fairly early stages, considerably increased tonnages of plastics will be used over the next decade.

There are some applications which are showing practically no increase or even a decline in some countries; one of these is the use of low tunnels in highly developed countries such as the USA, because there is a high labour content in operating them and the costs involved are not justified.

The handling of produce becomes more and more systematised as distribution areas take in whole continents instead of just the local markets and calls for properly designed containers not only to ensure safe transit but also to make most effective use of the transport capacity so as to keep the carrying costs to a minimum. This means that some measure of standardisation is essential and this can be most readily achieved by the use of moulded crates and trays.

There is every reason to suppose that the plastics and agricultural industries will be aware of the opportunities which exist and the advantages which can accrue from close collaboration on the improvement of existing techniques and the development of new ones.

USAGE OF PLASTICS IN NEW TECHNIQUES

The cost of energy for heating greenhouses for forced cultivation has become so high over recent years that many workers have been studying means of making savings and looking for heat sources which have up to now been neglected. In the first case, workers have been quick to take advantage of the cheapness and ease of handling of plastics to incorporate thermal insulation by use of double-walled structures and other techniques which have already been described.

The studies of heat sources which do not rely on fossil fuels have been concentrated on the utilisation of waste heat, primarily from industrial operations, and the capture and storage of solar energy. Solar captors are already well known but the cost of the installations which are commercially available is high and the investment would be excessive for growers wishing to heat their greenhouses. Recent studies have indicated much cheaper methods which are specially designed for horticulture and which utilise plastics materials.

Solar Captors

There are several possibilities by which solar energy can be trapped but there are two major classifications of solar collectors: concentrating and flat plate. The former can involve reflective mirrors that concentrate the sunlight onto a line or a point; most of these devices must be moved to a certain extent to track the sun and in addition their performance is quite low on hazy days.

Flat plate collectors are flat sheets of glass or plastics through which the sunlight passes; the sunshine is absorbed by the dark surface in which a calorific liquid circulates, and is converted to heat. Most collectors face south; if the tilt of the collector is equal to the latitude or up to the latitude plus 10°, the collection of heat will be similar during summer and winter. At the present time the collectors require excessive space for proper operation and are costly with high maintenance costs.

A system which has been developed at Rutgers University in the USA appears very promising and is relatively cheap compared with more sophisticated models.[214] The solar captor consists of four layers of clear film formed from two inflated sections with a black film sandwiched between them. The warm water heated by the solar collector is stored in the floor, which is formed by a layer of porous concrete on top of gravel, and this acts as thermal storage and as a heat exchanger. The water is retained by a PVC liner under the floor and heat loss into the soil is prevented by insulation with expanded polystyrene. This technique has been used in a greenhouse complex and appears to have considerable potential for reducing reliance on conventional heating procedures.

Solar Greenhouses

Recent research by the Institut National de la Recherche Agronomique in collaboration with the Centre d'Études Nucléaires (CEN) at Grenoble has led to the development of a greenhouse which obtains part of its fuel needs from solar energy. The essence of this solar greenhouse is a double translucent roof structure through which an aqueous solution of cupric chloride is circulated and this acts as a heat carrier. This solution is transparent to visible wavelengths and has a strong absorption coefficient in the near infra-red range. During the day the process produces good conditions for plant growth and reduces evaporation and ventilation needs. During the night the excess heat stored during the day in a reservoir is redistributed through the walls to help maintain the night temperature.

The solar energy collection efficiency of the greenhouse is within the range 30–40%; this heat represents approximately 40% of the night-time heat requirement. Preliminary tests have been reported to determine the optimum concentration of cupric chloride according to biological needs. This selective filter improves the physiological response of crops by reducing the plant temperature and the water requirements.

This type of structure appears to have considerable potential in those regions where high luminosity can allow some slight reduction of the transmission of visible radiation through the wall.

Other Potential Developments

The *heat pump* is one of the possible solutions being examined as a means of supplying cheaper energy for the greenhouse; an economic application is the use of the pump for watering with warm water.

It is often necessary to have large installations to supply a few cubic metres of warm water, but at the CEN at Grenoble an effective solution has been developed; this is a small electric heat pump which consumes only 2·4 kW/h. With 4 cm of water per hour at 23 °C, there is available 2 cm of water at 23 °C and this corresponds to two waterings. This system is already in use but the present price could be a drawback.

Considerable effort is being expended in the USA on solar thermal agricultural research programmes and it is likely that any solutions will utilise plastics materials. The five major areas being investigated are: grain drying, crop drying, greenhouses and rural residences, animal shelters and agricultural food processing.

References and Bibliography

Many of the references given in this section are to be found in publications concerned with the International Organisation for Plastics in Agriculture which is responsible for publishing a quarterly journal and for arranging the International Colloquia. The quarterly journal *Plasticulture* is published in two languages, English and French, while the proceedings of the international meetings are published in the language in which the original paper was given.

Photocopies of any of the references contained in the above publications can be obtained directly from: Comité International des Plastiques en Agriculture (CIPA), 18, place Henri Bergson, 75008 Paris, France.

The following are the abbreviations used for the publications which appear most frequently.

Plast.	*Plasticulture*, published by the CIPA.
4th Int. Coll. 1970	The proceedings of the International Colloquia are published by the national organisations in which country they are held, but copies are available from the CIPA at the address provided.
5th Int. Coll. 1972	
6th Int. Coll. 1974	
P.H.M.	*Pépinièristes, Horticulteurs, Maraichers*, 59 rue du Faubourg Poissonière, Paris 75009, France.
Plastu.	*Plasturgie*, Dubois, P. Published by Masson, 120, boul. St. Germain, Paris 6, France.
Plasto.	*Plastophysicochimie*, Dubois, P. Published by Masson, 120 boul. St. Germain, Paris 6, France.
G.P.A.	*Guide de l'Utilisateur des Plastiques en Agriculture*, published by the CIPA, 18, place Henri Bergson, 75008 Paris, France.
Techn. Ing.	*Techniques de L'Ingénieur*, published by Service Commercial, 123, rue d'Alesia, 75014 Paris, France.

REFERENCES

1. Manescu, B., *5th Int. Coll.*, 1972, p. 305.
2. Hamadi, M., *ibid.*, p. 489.
3. Bry, A., *ibid.*, p. 103.
4. Takahashi, K., *Plast.*, **15**, Sept., 1972, p. 36.
5. INRA, *Plast.*, **12**, Dec., 1971, p. 24.
6. De Neira, A. L., *Plasticos*, **17**(100), 1969, p. 92 (in Spanish).
7. Manescu, B., Sociu, S. and Lunca, L., *Int. Soc. Hort. Sci.*, Turin, 1967, Technical communication 9, The Hague, May, 1968, p. 95.
8. Duncan, G. A. and Walker, J., *Plast.*, 21 March, 1974, p. 4.
9. Keveren, R. I., *Plastics in Horticultural Structures*, Chs. 1 and 2, Rubber and Plastics Res. Ass., Shawbury, UK, 1976.
10. Pabiot, J., *Techn. Ing.*, A 3510–A 3512.
11. Reid, D. R., *Modern Packaging Films*, Ch. 7. Butterworths, London, 1967.
12. Davis, A. and Head, B. C., Paper C.1, *International Symposium, The Weathering of Plastics and Rubber*, Plastics and Rubber Inst., London, June 1976.
13. Cope, R. and Eurin, P., *ibid.*, Paper D.1.
14. Marechal, J. C. and Eurin, P., *ibid.*, Paper D.7.
15. Hanras, J., *ibid.*, Paper F.4.
16. Verdu, J., *Techn. Ing.*, A 3510–A 3151.
17. Verdu, J., *ibid.*, A 3162.
18. Winslow, F. H. and Hawkins, W. L., *Appl. Polymer Symposium*, No. 4, John Wiley & Sons, New York, 1967.
19. Goldsberry, K. L., *Colorado Flower Growers' Association Bulletin*, **202**, Feb., 1967.
20. Rosman, J., *Revue Belge des Matières Plastiques*, **5**(4), Aug., 1964, p. 365.
21. Wilson, D. D., *Proceedings, 10th National Agricultural Plastics Conf.*, Chicago, Nov., 1971, p. 160.
22. Davis, A., Deane, G. H. W. and Ledbury, J. K., Paper B.1, *International Symposium, The Weathering of Plastics and Rubber*, Plastics and Rubber Inst., London, June, 1976.
23. Verdu, J., *Techn. Ing.*, A 3151.
24. Robledo, F., *4th Int. Coll.*, 1970, p. 23.
25. Hanras, J. and Cluzeaud, A., *P.H.M.*, April, 1967.
26. Delano, C. J., *Plast.*, 14 June, 1972, p. 29.
27. Dubois, P., *Plastu.*, p. 35.
28. Buclon, F., *Plast.*, 11 Sept., 1971, p. 17.
29. Brighton, C. A., *5th Int. Coll.*, 1972, p. 203.
30. Langren, B., *ibid.*, p. 233.
31. Maquin, M., *Plast.*, **18**, June, 1973, p. 38.
32. Buttrey, D., *Mod. Plast. Int.*, **2**(2), Feb., 1972, p. 9.
33. Gilead, D., *Plast.*, **30**, June, 1976, p. 9.
34. Gilead, D., *ibid.*, **32**, Dec., 1976, p. 24.
35. Nissan Chem. Co., *Plastics and Rubber Weekly*, No. 659 (19 Nov., 1976), p. 11.

36. Favilli, R., Glatti, F. and Zanella, A., *Materie Plastiche ed Elastomeri*, **34**(6), June, 1968, p. 667 (in Italian).
37. Favilli, R., Glatti, F. and Zanella, A., *ibid.*, **34**(7), July, 1968, p. 799 (in Italian).
38. Favilli, R., Glatti, F. and Zanella, A., *ibid.*, **34**(8), Aug., 1968, p. 925 (in Italian).
39. Glatti, F., *4th Int. Coll.*, 1970, p. 205.
40. Aimi, R., *ibid.*, p. 197.
41. *La Plasticulture*, CdF Chimie Tour Aurore, La Defense, Paris, p. 28.
42. Yanagesawa, M. and Tezyka, Y., *Plast.*, **23**, Sept., 1974, p. 35.
43. Verdu, J., *Techn. Ing.*, A 3160.
44. *G.P.A.*, p. 146.
45. Freeman, B., *Plast.*, **32**, Dec., 1976, p. 45.
46. Gary, T., *Western Fruit Grower*, **14**(1), 1960, p. 34.
47. Spice, H. R., *Plast.*, **27**, Sept., 1975, p. 49.
48. Guyot, G., Thesis for Doctorate, Station Bioclimatologique, Montflavet, France, 1972.
49. Guyot, G., *10th Colloquium on Plastics in Agriculture*, Angers, June, 1967, p. 45.
50. Guyot, G., *4th Int. Coll.*, 1970, p. 294.
51. Quentin, M., *Plast.*, **16**, Dec., 1972, p. 23.
52. Mendez, L. A., *5th Int. Coll.*, 1972, p. 973.
53. Lafon, J. and Gouvernet, R., *Plast.*, **22**, June, 1974, p. 51.
54. Wood, S., *Mod. Plast.*, **46**(1), Jan., 1969, p. 10.
55. Anon., *Mod. Plast.*, **47**(1), Jan., 1970, p. 64.
56. Buclon, F., *5th Int. Coll.*, 1972, p. 5.
57. Buttrey, D. N., *Plastics and Polymers*, **37**(127), Feb. 1969, p. 65.
58. *G.P.A.*, p. 30.
59. Buclon, F., *Plast.*, **10**, June, 1971, p. 13.
60. Courter, J. W., Hopen, H. J. and Vandemark, J. S., *Proc. 8th National Agricultural Plastics Conference*, Univ. California, 1968, p. 126.
61. Abregts, E. E. and Howard, C. M., *HortScience*, **8**, 1973, p. 36.
62. Nieuwhof, B. J., *Plastics in Agriculture and Horticulture Conf.*, Wye College, March, 1974, Plastics and Rubber Inst., London.
63. Smith, N. J., *Plastics in Agriculture and Horticulture Conf.*, Wye College, March, 1974, Plastics and Rubber Inst., London.
64. Smith, N. J., *Am. Veg. Grower*, **21**(12), p. 13.
65. Smith, N. J., *Proceedings*, *11th National Agricultural Plastics Conference*, Univ. Texas, 1971, p. 167.
66. Charrier, M., *4th Int. Coll.*, 1970, p. 98.
67. Marcellin, P., *5th Int. Coll.*, 1972, p. 642.
68. Scott, K. J., *ibid.*, p. 729.
69. Bessemer, S. T., *6th Int. Coll.*, 1974, p. 263.
70. Riczko, J., *ibid.*, p. 1025.
71. *G.P.A.*, p. 99.
72. Ebel, R., *Plast.*, **21**, March, 1974, p. 51.
73. Quentin, M., *L'Irrigation et le Drainage en France*, Ste. Méditeraniene des Plastiques Agricoles (M.P.A.), CdF Chimie, Paris, 1974.

REFERENCES AND BIBLIOGRAPHY

74. Spice, H. R., *6th Int. Coll.*, 1974, p. 299.
75. Mendiz, L. A., *5th Int. Coll.*, 1972, p. 973.
76. Hall, B., *Plast.*, **29,** March, 1973, p. 39.
77. Peregi, S., *5th Int. Coll.*, 1972, p. 1029.
78. Tantini, L., *ibid.*, p. 1038.
79. Kehrer, K. K., Prefabricated shell structure of GRP for a greenhouse, Paper 26, *Conference on Plastics in Building Structures*, The Plastics Institute, London, 1965.
80. Vadasz, E., *5th Int. Coll.*, 1972, p. 1057.
81. Ronay, D. and Polgar, J., *ibid.*, p. 1064.
82. Ronay, D. and Falaky, J., *ibid.*, p. 1068.
83. Brighton, C. A., *Nurseryman and Garden Centre*, 27 Sept., 1973, p. 392.
84. Werminghausen, B., *Plast.*, **30,** June, 1976, p. 17.
85. Baumann, H., *5th Int. Coll.*, 1972, p. 929.
86. *La Plasticulture* CdF Chimie Tour Aurore, La Defense, Paris, p. 79.
87. Grobbelaar, H. L., *Plast.*, **16,** Dec., 1972, p. 27.
88. Chapin, R. D., *Proceedings, International Experts Panel on Irrigation*, Sept., 1971, Israel, pp. 1–8.
89. Buclon, F., *Plast.*, **21,** March, 1974, p. 35.
90. Koransky, Z. S. and Kiss, A. S., *5th Int. Coll.*, 1972, p. 899.
91. Hall, B. J., *Proceedings, 10th National Agricultural Plastics Conference*, Chicago, Nov., 1971, p. 69.
92. Laberche, J.-C., *Plast.*, **30,** June, 1976, p. 43.
93. Westwood, V., *Plast.*, **33,** March, 1977, p. 23.
94. Davids, G., *Drip/Trickle Irrigation*, **2**(2), April, 1977, p. 20.
95. Spice, H. R., *Plast.*, **27,** Sept., 1975, p. 49.
96. Quintard, A. M., *Plast.*, **26,** June, 1975, p. 65.
97. Sharp, H. E., *5th Int. Coll.*, 1972, p. 721.
98. Dubois, P., *Plasto.*, pp. 493–500.
99. Nielsen, L. E., *Mechanical Properties of Polymers*, Chapman & Hall, London, 1962.
100. Macskasy, H. and Szabo, A., *4th Int. Coll.*, 1970, p. 247.
101. Doc. CEMP 37/3, Centre d'Étude des Matières Plastiques, rue de Prony, Paris 17, France, June, 1974.
102. BAHPA Standard 0001:1976, British Agricultural and Horticultural Plastics Ass., 47 Piccadilly, London, W1, 1976.
103. Spice, H. R., *Plast.*, **18,** June, 1973, p. 43.
104. Skeist, I., *Handbook of Adhesives*, Reinhold, New York, 1962.
105. Ultrasonic fabrication, in *Encyclopaedia of Polymer Science and Technology*, Vol. 14, John Wiley & Sons, New York, 1971.
106. Spice, H. R., *Polyethylene Film in Horticulture*, Ch. 4, Faber & Faber, London, 1959.
107. 'Visqueen', *Polyethylene Films for Horticulture*, ICI Welwyn Garden City, p. 7.
108. Bugalho Semedo, C. M., *Plast.*, **27,** Sept., 1975, p. 22.
109. *La Plasticulture*, CdF Chimie Tour Aurore, La Defense, Paris, p. 99.
110. Bugalho Semedo, C. M., *5th Int. Coll.*, 1972, p. 606.
111. Cooper, A. L., *Scientia Horticulturae*, **3,** 1975, p. 25.

112. Spice, H. R., *Polyethylene Film in Horticulture*, Faber & Faber, London, 1959, p. 131.
113. Keveren, R. I., *Plastics in Horticultural Structures*, Rubber and Plastics Res. Ass., Shawbury, UK, 1976, pp. 164, 175, 180, 190.
114. Agulhon, R., *Plast.*, **11**, Sept., 1971, p. 39.
115. Agulhon, R., *5th Int. Coll.*, 1972, p. 577.
116. Agulhon, R., *6th Int. Coll.*, 1974, p. 177.
117. Agulhon, R., *Les Plastiques en Agriculture*, Comptes Rendues du Colloque National, 1976 (CIPA, Paris).
118. Hanras, J., *Plast.*, **14**, June, 1972, p. 18.
119. Garcia, V., *Fruits*, **23**(9), 1968.
120. Charpentier, J. M. et al., *Fruits*, **25**(2), 1970.
121. Monjardino, R., *5th Int. Coll.*, 1972, p. 620.
122. Ballif, J. L. and Dutil, P., *Plast.*, **16**, Dec., 1972, p. 33.
123. Ballif, J. L. and Dutil, P., *Plast.*, **22**, June, 1974, p. 7.
124. Buclon, F., *Plast.*, **21**, March, 1974, p. 35.
125. Dauple, M., *ibid.*, **26**, June, 1975, p. 25.
126. Schirmer, M., *ibid.*, **26**, June, 1975, p. 17.
127. Voth, V., *ibid.*, **29**, March, 1976, p. 15.
128. Voth, V., *ibid.*, **34**, June, 1977, p. 11.
129. Lang, M., *ibid.*, **34**, June, 1977, p. 23.
130. Emmert, E. M., Low cost plastics greenhouses, *Progress Report No. 28*, Agricultural Experimental Station, Univ. of Kentucky, June, 1975.
131. Lee Valley Experimental Horticulture Station, Film Plastic Tunnels, *Leaflet No. 17*, Hoddesdon, Herts., revised 1973.
132. Lee Valley Experimental Horticulture Station, Film plastic multispan greenhouses, *Leaflet No. 20*, Hoddesdon, Herts., revised 1973.
133. Spice, H. R., *Plastics*, **24**(263), Sept., 1959, p. 322.
134. Buttrey, D. N. and Spice, H. R., *Plastics Today*, **41**, 1971, p. 13, ICI, Welwyn Garden City.
135. Bry, A., *Plast.*, **12**, Dec., 1971, p. 15.
136. Hutton, B., *Gardeners' Chronicle/Horticultural Trade Journal*, **173**(12), p. 16.
137. Cotter, D. J. and Walker, J. N., *Proc. Am. Society of Horticultural Science*, **89**, 1966, p. 584.
138. Bartoli, R. and Gac, A., *5th Int. Coll.*, 1972, p. 288.
139. Keveren, R. I., *Plastics in Horticultural Structures*, Ch. 5, Rubber and Plastics Res. Ass., Shawbury, UK, 1976.
140. Sheldrake, R. and Langhans, R. W., Heating study with plastic greenhouses, *Proc. National Horticulture Plastics Conference*, Oct., 1961.
141. Roberts, W. J., *Proc. 8th National Agricultural Plastics Conference*, Univ. California, 1968.
142. Roberts, W. J. and Mears, D. R., *Trans. Am. Soc. of Ag. Engineers*, **12**(1), 1969, p. 32.
143. Besemer, S. T., Axlund, D. S. and Brown, A., *6th Int. Coll.*, 1974, p. 287.
144. Canham, A. E., Paper 1/F/12, Air supported plastics structures—materials and design factors, *Proc. Agricultural Engineering Symposium*, Silsoe, Sept., 1967.

145. Canham, A. E., *Int. Soc. Hort. Sci., Symposium*, Turin, 1967. Technical communication 9, The Hague, May, 1968, p. 105.
146. Canham, A. E., *Plastics and Polymers*, **37**(Bo. 130), Aug., 1969, p. 293.
147. D'Ogny, F., *Int. Soc. Hort. Sci. Symposium*, Turin, 1967. Technical communication 9, The Hague, May, 1968, p. 111.
148. Roberts, W. J., *Proc. 10th National Agricultural Plastics Conference*, Chicago, Nov., 1971, p. 103.
149. Bleasdale, J. K. A. and Thompson, R., Air Inflated Plastics Greenhouse, *Report of the National Vegetable Research Station*, 1968, p. 62.
150. Slann, P. A., Air-houses—some practical modern developments, *Farm Mechanisation and Buildings*, April, 1968.
151. Cropping in walk-in film plastic tunnels, *Leaflet No. 21*, Lee Valley Experimental Horticulture Station.
152. Skierkowski, J., *5th Int. Coll.*, 1972, p. 479.
153. Benoit, F., *ibid.*, p. 398.
154. Benoit, F., *ibid.*, p. 407.
155. Guariento, M. and Ravilli, V., *ibid.*, p. 416.
156. Loche, M., *Plast.*, **20**, Dec., 1973, p. 41.
157. Jasa, B., *5th Int. Coll.*, 1972, p. 432.
158. Benoit, F., *ibid.*, p. 411.
159. Harkavi, Y., *Plast.*, **34**, June, 1977, p. 39.
160. Rodeyns, A., *Revue Belge des Matières Plastiques*, **5**(4), Aug., 1964, p. 307.
161. Goldsberry, K. L., *Proc. 10th National Agricultural Plastics Conference*, Chicago, Nov., 1971, p. 154.
162. Harnett, R. F., *The Grower*, 14 Oct., 1972.
163. Buttrey, D. N., Plastics in agriculture and horticulture, *Plastics Today*, **47**, March, 1974, p. 2, ICI, Welwyn Garden City, UK.
164. Wilson, D. D., *Proc. 10th National Agricultural Plastics Conference*, Chicago, Nov., 1971, p. 160.
165. Keveren, R. I., *Plastics in Horticultural Structures*, Rubber and Plastics Res. Ass., Shawbury, UK.
166. Buttrey, D. N., *4th Int. Coll.*, 1970, p. 21.
167. Werminghausen, B., *Kunststoffe*, **54**(10), Oct., 1964, p. 391.
168. Werminghausen, B., *ibid.*, **55**(5), May, 1965, p. 391.
169. Ruthner, O., *ibid.*, **55**(12), Dec., 1965, p. 23.
170. Buclon, F., *Plast.*, **27**, Sept., 1975, p. 33.
171. Dumont, M., Dalle, M. Y. and Giannesini, J.-P., *7th Int. Coll.*, San Diego, 1977.
172. ADAS Liaison Unit, National Institute of Agricultural Engineering, *The Grower*, 13 Feb., 1971.
173. Martineau, F., *Connaissance des Plastiques*, **5**(50), Dec., 1974, p. 37.
174. Walker, J. N. and Cotter, D. *Materie Plastiche ed Elastomeri*, **32**(3), March, 1966, p. 321.
175. Buclon, F., *Int. Soc. Hort. Sci. Symposium*, Turin, 1967. Technical Communication 9, The Hague, May, 1968, p. 31.
176. Hall, B. J., *Proc. 10th National Agricultural Plastics Conference*, Chicago, Nov., 1971, p. 131.
177. Werminghausen, B., *Plast.*, **12**, Dec., 1972, p. 5.

178. Delano, R. and Raseman, C. J., Control of condensate and light in greenhouses and solar stills, *Plastics in Agriculture and Horticulture Conf.*, Wye College, March, 1974, Plastics and Rubber Inst., London.
179. Buclon, F., *Plast.*, **21,** March, 1974, p. 35.
180. Dezo, D., *5th Int. Coll.*, 1972, p. 1017.
181. Papenduk, K., *Plast.*, **33,** March, 1977, p. 37.
182. Hyde, M., *4th Int. Coll.*, 1970, p. 183.
183. Walley, R. P., *5th Int. Coll.*, 1972, p. 993.
184. Hall, B. J., *6th Int. Coll.*, 1974, p. 167.
185. Voth, V., Munnecke, D. E. and Paulus, A. O., *ibid.*, p. 173.
186. Water storage, *Technical Leaflet*, ICI, Welwyn Garden City, UK.
187. Spice, H. R., *5th Int. Coll.*, 1972, p. 964.
188. Staff, C. E., *Plast.*, **14,** June, 1972, p. 22.
189. Brent Cluff, C., *Proc. 10th National Agricultural Plastics Conference*, Chicago, Nov., 1971, p. 193.
190. Borca, B., *Plast.*, **27,** Sept., 1975, p. 16.
191. Lemcoff, J., *5th Int. Coll.*, 1972, p. 878.
192. Sarathy, M. P., *Plast.*, **28,** Dec., 1975, p. 51.
193. Nisen, A., *5th Int. Coll.*, 1972, p. 194.
194. Damagnez, H. and Chiapale, J. P., *Plastics in Agriculture and Horticulture Conf.*, Wye College, March, 1974, Plastics and Rubber Inst., London.
195. Bry, A. and Daverat-Julian, L., *Les Plastiques en Agriculture*, Comptes Rendues du Colloque National, 1976 (CIPA, Paris), p. 91.
196. Stickler, P., *Plast.*, **25,** March, 1975, p. 41.
197. Cooper, A. L., *The Grower*, 5 May, 1973.
198. Cooper, A. L., *ibid.*, 2 March, 1974.
199. Cooper, A. L., *ibid.*, 25 Jan., 1975.
200. Morgan, J. L., The prevention of wind erosion of light soils, *Plastics in Agriculture and Horticulture Conf.*, Wye College, March, 1974, Plastics and Rubber Inst., London.
201. Staudinger, J. J. P., Plastics Waste and Litter, *SCI Monograph No. 35*, Society of Chemical Industry London, 1970.
202. Tuyaux en polyethylene—basse densité, CEMP 02/6 May, 1971.
203. Tuyaux en polyethylene—haute densité, CEMP 04/2 Jan., 1972.
204. Low density polyethylene pipe, BS 1972: 1967.
205. High density polyethylene pipe, BS 3284: 1967.
206. Tubes et raccords en PVC non-plastifié, CEMP 36/9 May, 1976.
207. UPVC pipe for cold water services, BS 3505: 1968.
208. UPVC pipe for industrial use, BS 3506: 1968.
209. Joints and fittings for use with UPVC pressure pipes, BS 4346 Part 1: 1969.
210. Garnaud, J. C., *Plast.*, **19,** Sept., 1973, p. 7.
211. Damagnez, J., Samie, C., de Villele, O. and Dauple, P., *5th Int. Coll.*, 1972, p. 356.
212. Brun, R. and Laberche, J.-C., *Les Plastiques en Agriculture*, Comptes Rendues du Colloque National, 1976 (CIPA, Paris), p. 23.
213. Jensen, M. H., *6th Int. Coll.*, 1974, p. 307.
214. Roberts, W. J. and Mears, D. R., *Plast.*, **33,** March, 1977, p. 29.

FURTHER READING

A global view of greenhouse food production, *Report No.* 89, Economic Research Service, US Dept. of Agriculture, Washington D.C., Oct., 1973.

Allington, P. and Allen, P. G., Film plastic greenhouses for horticultural crops, *Agriculture*, **79**(3), March, 1972, p. 109.

Buclon, F., Plasticulture in the world. Paper presented to *International Conference*, Baghdad, 1975 (available from CIPA).

Eaton, H. J., Shelter belts and hedges, *Agriculture*, **78**(5), May, 1971.

Mullins, D. H., Cooper, C. C. and Clarkson, V. A., Horticultural and agricultural applications in *Encyclopaedia of Polymer Science and Technology*, Vol. 7, John Wiley & Sons, New York, 1976.

Sheard, G. F., Shelter and the effect of wind on the heat loss from greenhouses, *Acta Horticultura*, **76**, 1977.

Spice, H. R., The use of plastics for protected environments, *Outlook on Agriculture*, **8**(2), 1974, p. 63.

Vegetable crop responses to synthetic mulches, *NAPA Tech. Bulletin* 1, National Agricultural Plastics Association.

Index

Adhesion, 77
  heterogeneous, 77
Adhesive emulsion, 77
Adhesive paint, 81
Adhesives, 77
Ageing, 54, 73
  artificial, 74
  factors affecting, 73–4
  resistance to, 84
Animal protection, 68
Applications, 9
Arrhenius Law, 17
Atmosphere control in greenhouses, 135

Balanced film, 71
Baler twine, 68
Bananas, mulching, 85
Boxes, 64
British Agricultural and Horticultural Plastics Association (BAHPA)
  standard, 75
  test, 75
Buildings, 47

Canals, 63, 130
Carbon dioxide
  atmosphere, in, 2, 47
  exchange of, 38
Carrots, greenhouse cultivation, 102
Centre d'Étude des Matieres Plastiques (CEMP) standard, 75
Channels, 130

Chemical resistance of thermoplastics, 32
Chlorophyll
  action of, 2
  forms of, 2
Citrus fruits, mulching, 85
Clay tiles and pipes, 63
Climate, 6
  classification, 6
  see also Microclimate
Cocoa, mulching, 85
Coffee, mulching, 85
Combustibility, 28
Containers, 66
Crates, 64
Crop-gathering, 42
Crop rotation, 52–3
Cultivation techniques, 149–51
Cuttings, 138

Dams, 129
Degree days, 20
Degree of stretching, 69
Density of films, 92
Desiccation, 45
Dormancy in plants, 60
Double-film structure, 48
Drainage and drains, 60, 62, 63, 131
  deep, 61
  expanded plastics, 133
  laying, 131
  surface, 60
  water table between, 61
Drying, 45

Ecology, 140
Economics, 152
  greenhouses, of, 135
Electrical conductivity, 54
Electrical insulants, 18
Electrostatic charges, 18–20, 54
Elongation, 71, 74
Embrittlement, 74, 75
'Empot' handling system, 63
Endive, greenhouse cultivation, 103
Energy
  input, 17–28
  required for fracture, 32
  saving, 136
Equipment, 63
Ethylene–vinyl acetate (EVA), 11, 88, 150
Evaporation in various regions, 52
Evapotranspiration, 38, 49
Expansion coefficient, 69
Extrusion, 72
  blowing, 70

Fibre filament, 69
Film(s), 1, 43–53, 65, 150
  anti-mist, 24
  'balanced', 71
  black, specification for, 145
  contact embrittlement, 75
  cutting, 76
  density, 92
  extrusion blowing, 70
  fixing, 79
  functions, 43–53
  gas and water permeability, 29
  manufacturing supervision and control, 143
  marking of, 143
  perforated, 91
  permeable, 91
  photo-degradable, 25
  physical properties after use, 20
  silage, 45, 120, 143
  sleeves, 68
  soil temperature under, 26
  special, 24
  standard tests for, 143

Film(s)—*contd.*
  standards, 145
  thermal, 24
  transmittance, 20
  UV stabilised or long-life, 24
  see also Greenhouses; Mulching; Tunnel structures
Fixing, 79
Fracture, energy required for, 32
Frost protection, 40–1, 58
Fumigation, 52–3

Gluing, 77
Grafts, 138
Granules, 54
Greenhouse(s), 92–116, 150
  air-inflated, 39
  air-supported, 98
  arched or curved roofs, 108
  atmosphere control, 135
  condensation, 116
  construction of, 135
  cultivation, 101–5
  development in UK, 99–101
  double-layer, 96–8, 136
  economics, 135
  effect, 22, 23
  energy saving, 136
  film properties, 92
  glazed with rigid sheets, 151
  heated, 151
  heating, 110–13
  inflatable, 136
  insulation, 137
  irrigation, 59
  lining, 115–16
  metal frameworks, 96
  mobile, 108
  polyethylene film for, 145
  reflector screens, 136
  reinforced PVC tunnels and multi-span structures, 151
  rigid, 105–9
  self-supporting, 108
  solar, 160
  temperature, 116
  thermal spar, 136

INDEX

Greenhouse(s)—*contd.*
  tower, 108
  traditional form, 107
  tunnel, 113
  types of, 95
  unheated, 151
  ventilation, 93, 113–15
  wooden frameworks, 96
Growth retardation, 150

Hail
  protection against, 42
  resistance, 95
Handling, 64
Heat
  absorption, 48
  pump, 161
  transfer, 37
Heating
  air–soil, 111
  flexible tube soil, 111
  greenhouses, 93, 110–13
  steam or hot water, 110
  warm air, 110
Hill lakes, 126
Humidity, 8, 135
  greenhouses, of, 93
Hydroponic cultivation, 63, 137–8

Identification of plastics, 148
Irrigation, 55, 62, 122
  channel, 55
  distribution systems, 62
  greenhouses, 59
  pipe network, 57
  spray, 57
  trickle or drip, 58

Joining, 79

'Key-Emitter' micro-tube system, 125

Labelling, 81

Layering, 138
Lettuce, greenhouse cultivation, 102
Liners, 53–4
Low tunnels, 89, 105, 134, 150

Machinery, 63
Maize, mulching, 85
Mass transfer, 38
Materials, 1, 54
Mechanical properties of polymers, 146
Mechanisation, 63
Melons
  greenhouse cultivation, 101
  mulching, 86
Microclimate, 6, 43, 88
Micro-tube drippers, 122–6
  bi-wall, 122
  drip-head, 125
  perforated tubes, 126
  porous tube (Viaflo), 124
  twin-wall, 122
Mineral glasses general properties, 32
Molecular orientation, 69
Mole-plough and pipe layer, 132
Mulch films, 43, 44
Mulch trickle irrigation, 59
Mulching, 24, 43, 82–8, 117–19, 149, 150, 156
  cultivation, 84–8
  protective structures, in, 92
  vines, of, 83–4
Mushroom houses, 49

Net(s), 39–42
  functions, 40
  optimum properties of, 42
  protection of plants, 40
  supports for fungicide or insecticide, 42
Nitrogen cycle, 6
Nursery stock, mulching, 85
Nutrient film technique, 140

Organic glasses, general properties, 32

Orientation, 69, 70
Oxyazohydrocarbons (ONHC), 28

Packaging, 44, 64
Pears, mulching, 85
Permeability to liquids and gases, 29
Photochemical reactivity, 6
Photo-oxidation, 17
  degradation by, 73
Photoselectivity, 25
Photosynthesis, 2
Pipes, 1, 54, 61, 72, 122
  drainage, 62
  fittings, and, 80
  performance, 58
  polythene, 148
  standards tests for, 148
  working pressures, 148
Plant
  biology, 1
  dormancy, 60
Plastics, 9
  agriculture, in, 1
  breakdown of tonnage for 1976, 156
  compositions, 10
  extension of present uses, 159
  filled, 10
  foamed, 11
  identification of, 148
  plasticised, 10
  reinforced, 10
  usage
    France for cultural applications, in, 157
    Italy and Portugal, in, 157
    new techniques, in, 159
    UK, in, 157-8
    various countries for cultivation techniques, in, 155-6
Plasticulture
  definition, 1
  general background to, 9-32
  practice, in, 82
  prospects for, 155-61
  results achieved, 149-54
  world statistics, 155
  world-wide development, 153

Pollution, 140
Polyesters
  optical properties, 106
  reinforced, 105
Polyethylene, 11, 88, 92, 138, 150
  film
    durability of, 14
    greenhouses, for, 145
    UV-stabilised, 17
Polymers, 9
  cross-linked, 10
  linear, 9
  mechanical properties of, 146
  pure, 9
Polymethyl methacrylate (PMMA), 106
Polypropylene, 157
Polystyrene, expanded, 137
Polythene pipe, 148
Polyvinyl chloride, 11, 88, 150, 151
  reinforced, 133
  rigid, 106, 157
Polyvinyl fluoride, 17
Pots, 66
Powders, 54
Produce conservation, 44
Protected zone, 35
Protective sleeves, 68
Protective tunnels, 132-41

Rabbits, protection against, 140
Radiation transmission, 20, 22
Rainfall, 49
Reservoirs, 126
Resins, 10
Rodents, protection against, 140
Rods, 72
Rupture strength, 71

Sacks, 67
Semi-finished products, 69-81
Semi-forcing, 88, 105
Shading, 65
Shaping of sheets, 76
Sheets, 1, 47, 53-4, 64, 65, 72, 80
  corrugated, 80

INDEX

Sheets—*contd.*
  cutting, 76
  shaping, 76
Silage, 119–21, 157
  films, 45, 120, 143
  plastics film, under, 45
  vacuum, 120
Silo
  design, 47
  towers, 119
Soil
  aerators, 54
  erosion, prevention of, 140
  permeability, 60
  radiation emission by, 25
  sterilisation, 43, 52–3, 121–2
  temperature under film, 26
  water retention properties of, 51
Soil-less culture, 137–8
Solar captors, 160
Solar energy, 2, 17
Solar greenhouses, 160
Solar radiation, 22, 76, 93
Solar research, 161
Specific gravity, 27
Specific heat, 26
Specifications, 142–8
  black films, for, 145
Stack consolidation, 47
Standards and standard tests, 142–8
  films, for, 143, 145
  pipes, for, 148
Strawberries
  greenhouse cultivation, 103
  mulching, 87
Stress–strain diagrams, 72, 73
'Sun Clear', 23
Sunshine hours, 17
Surface drainage, 60
Sweet peppers, greenhouse cultivation, 103
Synthesis, 49

Tanks, 67
Tapes, 72
Temperature differences, 8
Tensile properties, 72–3

Thermal conductivity, 26
Thermal diffusivity, 27
Thermal expansion, 81
Thermal insulation, 26–8, 48, 54
Thermal transmission, 20
Thermoplastics, 10, 11
  ageing, 15
  chemical resistance of, 32
  elongation at break, 14
  impact resistance, 15
  modulus of elasticity, 11
  physical and transmission characteristics, 11
  properties of, 11–17
  rupture strength, 14
  test methods, 15
Thermosets, 10, 11
Tomatoes, 151
  greenhouse cultivation, 102, 104
  nutrient film technique, 139
Toxicity, 28
  vinyl chloride, of, 28
Trays, 63, 64
Trenching machine, 133
Tropical horticulture, 151–2
Troughs, 63
Tubing, 1, 45, 49, 54
  perforated layflat, 131
Tunnel structures, 88, 105
  films used in, 89
Turbulent zone, 33

Utilisable water reserve (UWR), 51

Vacuum method, 47
Ventilation of greenhouses, 93, 113
Vessels, 53–4, 68
Vines, mulching of, 83–4

Waste and waste problems, 140–1
Water
  catchment, 127
  demands, 55
  management, 122–33
  plasticulture, in, 49
  required by plants, 49

Water—*contd.*
  requirements, 52
  reserves, 126
  retention properties of soils, 51
  savings, 140
  supplies, 148
  table between drains, 61
  transfer, 38
Weed suppression, 84
Welding, 77–81
  high-frequency (dielectric), 78
  hot air torch, 78
  hot iron, 79
  infra-red, 78
  solvent, 80
  ultrasonic, 78

Wetting agent, 23, 24
Wind speed effect, 37
Windbreaks, 33–9
  biological effects, of, 38
  climate, and, 37
  effects of height and porosity of, 35
  impermeable, 33
  mechanisms of functioning of, 33
  permeable, 33, 35
  porosity of, 35
  reduction of heat transfer in vertical and horizontal directions, 37
  roughness of ground area, 37
  successive, 37
  wind speed effect, 37